STEEL DETAILERS' MANUAL

D

STEEL DETAILERS' MANUAL

ALAN HAYWARD CEng, FICE, FIStructE, MIHT

and

FRANK WEARE MSc(Eng), DIC, DMS, CEng, FIStructE, MICE, MIHT, MBIM

OXFORD

BLACKWELL SCIENTIFIC PUBLICATIONS

LONDON   EDINBURGH   BOSTON

MELBOURNE   PARIS   BERLIN   VIENNA

Blackwell Scientific Publications
Editorial Offices:
Osney Mead, Oxford OX2 0EL
25 John Street, London WC1N 2BL
23 Ainslie Place, Edinburgh EH3 6AJ
238 Main Street, Cambridge,
    Massachusetts 02142, USA
54 University Street, Carlton
    Victoria 3053, Australia

Other Editorial Offices:
Librairie Arnette SA
2, rue Casimir-Delavigne
75006 Paris
France

Blackwell Wissenschafts-Verlag
Meinekestrasse 4
D-1000 Berlin 15
Germany

Blackwell MZV
Feldgasse 13
A-1238 Wien
Austria

First published 1989

Reprinted 1990
This edition 1992

Set by DP Photosetting, Aylesbury, Bucks
Printed and bound in Great Britain by
Redwood Press Ltd, Melksham, Wiltshire

DISTRIBUTORS

Marston Book Services Ltd
PO Box 87
Oxford OX2 0DT
(*Orders:* Tel: 0865 791155
        Fax: 0865 791927
        Telex: 837515)

USA
    Blackwell Scientific Publications, Inc.
    238 Main Street
    Cambridge, MA 02142
    (*Orders:* Tel: 800 759-6102
            617 225-0401)

Canada
    Oxford University Press
    70 Wynford Drive
    Don Mills
    Ontario M3C 1J9
    (*Orders:* Tel: 416 441-2941)

Australia
    Blackwell Scientific Publications
    (Australia) Pty Ltd
    54 University Street
    Carlton, Victoria 3053
    (*Orders:* Tel: 03 347-0300)

British Library
Cataloguing in Publication Data
A Catalogue record for this book is
available from the British Library

ISBN 0-632-03523-4

Diagrams and details presented in this manual were prepared by structural draughtsmen employed in the offices of Cass Hayward and Partners, Consulting Engineers of Chepstow (Gwent, UK), who are regularly employed in the detailing of structural steelwork for a variety of clients including public utilities, major design: build contractors and structural steel fabricators.

Care has been taken to ensure that all data and information contained herein is accurate to the extent that it relates to either matters of fact or opinion at the time of publication. However neither the authors nor the publishers assume responsibility for any errors, misinterpretation of such data and information, or any loss or damage related to its use.

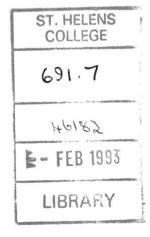

CONTENTS

LIST OF FIGURES

LIST OF TABLES

FOREWORD

BY THE DIRECTOR OF THE BRITISH
CONSTRUCTIONAL STEELWORK ASSOCIATION

Use of structural steel for building frames is steadily increasing compared with other materials, due in part to the demands for ever shorter construction periods, a challenge for which the material is ideally suited. The ability of steel to achieve a rapidly executed secure framework with flexibility for future use is now well known, allied with modern improvements in claddings and floor construction. Developments in the UK and European fabrication industry including use of automated techniques now enable the attributes to be fully utilised. The improved market share in buildings in being recognised in other fields, for example in foundation construction and for bridges. Therefore structural steel is likely to consolidate its position and continue as a vital medium in building the structures of tomorrow.

Authors Alan Hayward and Frank Weare have written the *Steel Detailer's Manual* as an education tool and to advance the knowledge of all those who may become involved with steel in construction by giving guidance on a much neglected aspect – that of *detailing*. They have rightly recognised that the viability and safety of steel structures relies on practical details which allow economical fabrication and final erection. The importance of appreciating and providing for tolerances so as to avoid unnecessary site rectification is given attention, and the examples of 'do's and don'ts' are a useful way of guiding detailers clear of some of the pitfalls which have previously caused problems and delays. Some of the implications of design upon detailing and fabrication and vice versa are covered, together with general information on the behaviour of steel as a material, often missing in other textbooks. Much of the data is available elsewhere, but perhaps for the first time the Authors have been able to draw detailing information together.

It is often the practice for engineers responsible for the structure to rely upon the fabricator manufacturing the steelwork to carry out the detailing including design of the connections. Increasingly the time being made available for this vital process is being reduced by shorter delivery periods, yet the safety of the steel frame relies upon the efficacy of the details and connections. Hopefully detailers and all other professionals involved in the construction process will use this book to improve their efficiency in rapid completion of steel frames from conception to occupation.

Derek Tordoff
Director
British Constructional Steelwork Association

May, 1988

PREFACE

The purpose of this manual is to provide an introduction and guide to draughtsmen, draughtswomen, technicians, structural engineers, architects and contractors in the detailing of structural steel. It will also provide a textbook for students of structural engineering, civil engineering and architecture. In addition, it will advise staff from any discipline who are involved in the fabrication and erection of steelwork, since the standards shown on the detailer's drawings have such a vital influence on the viability of the product and therefore the cost of the overall contract.

Detailing is introduced by describing common structural shapes in use and how these are joined to form members and complete structures. The importance of tolerances is emphasised and how these are incorporated into details so that the ruling dimensions are used and proper site fit-up is achieved. The vital role of good detailing in influencing construction costs is described. Detailing practice and conventions are given with the intent of achieving clear and unambiguous instructions to workshop staff. The authors' experience is that, even where quite minor errors occur at workshop floor or at site, the cost and time spent in rectification can have serious repercussions on a project. Some errors will, of course, be attributable to reasons other than the details shown in drawings, but many are often due to lack of clarity and to ambiguities in detailing, rather than actual mistakes. The authors' view is that the role of detailing is a vital part of the design: construction process which affects the success of steel structures more than any other factor. In writing this manual the authors have therefore striven to present the best standards of detailing allied to economic requirements and modern fabrication processes.

The concept of *engineer's drawings* and *workshop drawings* is described and the purpose of each explained in obtaining quotations, providing the engineer's requirements, and in fabricating the individual members. Detailing data and practice are given for standard sections, bolts and welds. Protective treatment is briefly described with typical systems tabled. Examples are given of arrangement and detail drawings of typical structures, together with brief descriptions of the particular design features. It is important that draughtsmen appreciate the design philosophies underlying the structure they are detailing. Equally the authors are concerned that the designer should fully appreciate details and their practical needs which may dictate design assumptions.

Examples of structures include single and multi-storey buildings, towers and bridges. Some of these are taken from actual projects, and although mainly designed to UK Codes, will provide a suitable basis for similar structures built elsewhere. This is because the practices used for steelwork design fabrication and erection are comparable in many countries of the world. Although variations will arise depending upon available steel grades, sizes, fabrication and erection plant, nevertheless methods generally established are adaptable, this being one of the attributes of the material. The examples show members sizes as actually used but it is emphasised that these might not always be suitable in a particular case, or under different Codes of Practice. Sizes shown however provide a good guide to the likely proportions and can act as an aid to preliminary design. Generally the detailing shown will be suitable in principle for fabrication and erection in many countries. Particular steelwork fabricators have their own manufacturing techniques or equipment, and it should be appreciated that details might need to be varied in a particular case.

The authors acknowledge the advice and help given to them in the preparation of this manual by their many friends and colleagues in construction. In particular thanks are due to the British Steel Corporation, the British Constructional Steelwork Association and the Steel Construction Institute who gave permission for use of data.

Alan Hayward
Cass Hayward & Partners
Consulting Engineers
York House, Welsh Street
Chepstow, Gwent NP6 5UW

Frank Weare
Polytechnic of Central London
35 Marylebone Road
London NW1 5LS

May 1988

1 USE OF STRUCTURAL STEEL

1.1 WHY STEEL?

Structural steel has distinct capabilities compared with other construction materials such as reinforced concrete, prestressed concrete, timber and brickwork. In most structures it is used in combination with other materials, the attributes of each combining to the whole. For example, a factory building will usually be steel framed with foundations ground and suspended floors of reinforced concrete. Wall cladding might be of brickwork with the roof clad with profiled steel or asbestos cement sheeting. Stability of the whole building usually relies upon the steel frame, sometimes aided by inherent stiffness of floors and cladding. The structural design and detailing of the building must consider this carefully and take into account intended sequences of construction and erection. Compared with other media, structural steel has attributes as given in Table 1.1.

In many projects the steel frame can be fabricated while the site construction of foundations is being carried out. Steel is also very suitable for phased construction which is a necessity on complex projects. This will often lead to a shorter construction period and an earlier completion date.

Steel is the most versatile of the traditional construction materials and the most reliable in terms of consistent quality. By its very nature it is also the strongest and may be used to span long distances with a relatively low self weight. Using modern techniques for corrosion protection the use of steel provides structures having a long reliable life, and allied with use of fewer internal columns achieves flexibility for future occupancies. Eventually when the useful life of the structure is over, the steelwork may be dismantled and realise a significant residual value not achieved with alternative materials. There are also many cases where steel frames have been used again re-erected elsewhere.

Structural steel can, in the form of composite construction co-operate with concrete to form members which exploit the advantages of both materials. The most frequent application is building floors or bridge decks where steel beams support and act compositely with a concrete slab via shear connectors attached to the top flange. The compressive capability of concrete is exploited to act as part of the beam upper flange, tension being resisted by the lower steel flange and web. This results in smaller deflections than those to be expected for non-composite members of similar cross-sectional dimensions. Economy results because of best use of the two materials – concrete which is effective in compression – and steel which is fully efficient when under tension. The principles of composite construction for beams are illustrated in figure 1.1 where the concept of stacked plates shown in (a) and (b) illustrates that much greater deflections occur when the plates are horizontal and slip between them can occur due to bending action. In composite construction relative slip is prevented by shear connectors which resist the horizontal shear created and which prevent any tendency of the slab to lift off the beam.

Structural steel is a material having very wide capabilities and is compatible with and can be joined to most other materials including plain concrete, reinforced or prestressed concrete, brickwork, timber, plastics and earthenware. Its co-efficient of thermal expansion is virtually identical with that of concrete so that differential movements from changes in temperature are not a serious consideration when these materials are combined. Steel is often in competition with other materials, particularly structural concrete. For some projects different contractors often compete to build the structural frame in steel or concrete to maximise use of their own particular skills and resources. This is healthy as a means of maintaining reasonable construction costs. Steel though is able to contribute effectively in almost any structural project to a significant extent.

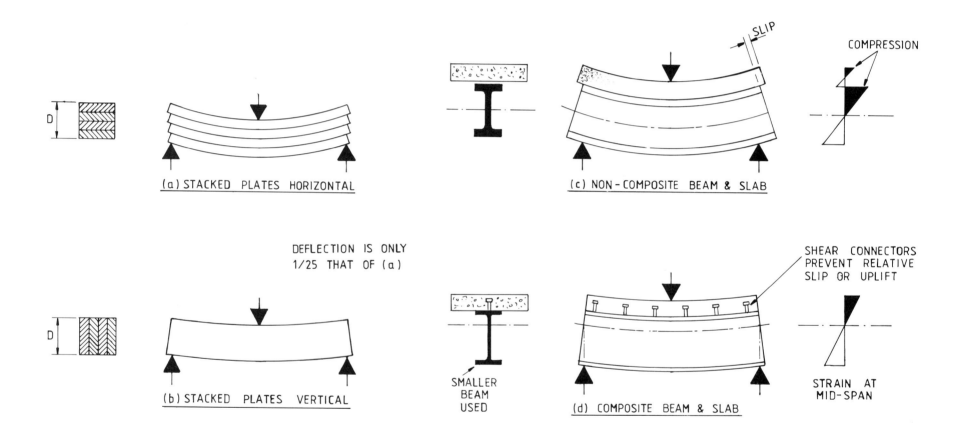

Figure 1.1 Principles of composite construction

Table 1.1 Advantages of structural steel

Feature	Leading to	Advantage	
		in buildings	in bridges
1 Speed of construction	Quick erection to full height of self supporting skeleton	Can be occupied sooner	Less disruption to public
2 Adaptability	Future extension	Flexible planning for future	Ability to upgrade for heavier loads
3 Low construction depth	Reduced height of structure	Cheaper heating Reduced environmental effect	Cheaper earthworks Slender appearance
4 Long spans	Fewer columns	Flexible occupancy	Cheaper foundations
5 Permanent slab formwork	Falsework eliminated	Finishes start sooner	Less disruption to public
6 Low weight of structure	Fewer piles and size of foundations Typical 50% weight reduction over concrete	Cheaper foundations and site costs	
7 Prefabrication in workshop	Quality control in good conditions avoiding sites affected by weather	More reliable product Fewer specialist site operatives needed	
8 Predictable maintenance costs	Commuted maintenance costs can be calculated. If repainting is made easy by good design, no other maintenance is necessary	Total life cost known Choice of colour	
9 Lightweight units for erection	Erection by smaller cranes	Reduced site costs	
10 Options for site joint locations	Easy to form assemblies from small components taken to remote sites	Flexible construction planning	

1.2 STRUCTURAL STEELS

1.2.1 Requirements
Steel for structural use is normally hot rolled from billets in the form of plate flat or section at a rolling mill by the steel producer, and then delivered to a steel fabricator's workshop where components are manufactured to precise form with connections for joining them together at site.

Frequently used sizes and grades are also supplied by the mills to steel stockholders from whom fabricators may conveniently purchase material at short notice, but often at higher cost. Fabrication involves operations of sawing, shearing, punching, grinding, bending, drilling and welding to the steel so that it must be suitable for undergoing these processes without detriment to its required properties. It must possess reliable and predictable strength so that structures may be safely designed to carry the specified loads. The cost : strength ratio must be as low as possible consistent with these requirements to achieve economy. Structural steel must possess sufficient ductility so as to give warning (by visible deflection) before collapse conditions are reached in any structure which becomes unintentionally loaded beyond its design capacity and to allow use of fabrication processes such as cold bending. The ductility of structural steel is a particular attribute which is exploited where the 'plastic' design method is used for continuous (or statically indeterminate) structures in which significant deformation of the structure is implicit at factored loading. Provided that restraint against buckling is ensured this enables a structure to carry greater predicted loadings compared with the 'elastic' approach (which limits the maximum capacity to when yield stress is first reached at the most highly stressed fibre). The greater capacity is achieved by redistribution of forces and stress in a continuous structure, and by the contribution of the entire cross section at yield stress to resist the applied bending. Ductility may be defined as the ability of the material to elongate (or strain) when stressed beyond its yield limit shown as the strain plateau in figure 1.2. Two measures of ductility are the 'elongation' (or total strain at fracture) and the ratio of ultimate strength to yield

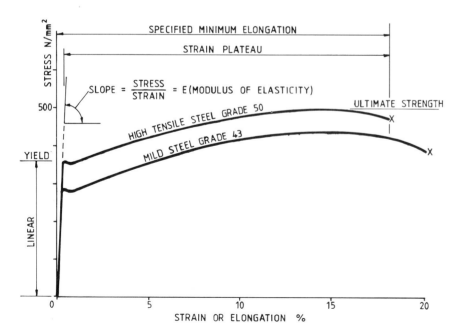

Figure 1.2 Stress: strain curves for structural steels

strength. For structural steels these values should be at least 18 per cent and 1.4 respectively.

For external structures in cold environments (i.e. typically in countries where temperatures less than about 0°C are experienced) then the phenomenon of *brittle fracture* must be guarded against. Brittle fracture will only occur if the following three situations are realised simultaneously:

(1)   A high tensile stress.
(2)   Low temperature.
(3)   A notch-like defect or other 'stress raiser' exists.

The stress raiser can be caused by an abrupt change in cross section, a weld discontinuity, or a rolled-in defect within the steel. Brittle fracture can be overcome by specifying a steel with known 'notch ductility' properties usually identified by the 'Charpy V-notch' impact test, measured in terms of energy in joules at the minimum temperature specified for the project location.

These requirements mean that structural steels need to be weldable low carbon type. In many countries a choice of mild steel or high tensile steel grades are available with comparable properties. In the UK structural steel is obtained to BS 4360 [1] in which the most commonly used grades are 43 (mild) and 50 (high tensile steel). The grades are further designated by a letter (e.g. 43A, 43B etc.) which denotes the requirements for Charpy V-notch impact testing. The letter 'A' requires no such testing with each letter starting from 'B' denoting an increased requirement (i.e. tested at a lower temperature). The main properties for the most commonly used grades are summarised in Table 1.2.

1.2.2 Recommended grades

In general it is economic to use high tensile steel (grade 50) due to its favourable cost : strength ratio compared with mild steel (grade 43) typically showing a 20% advantage. Where deflection limitations dictate a larger member size (such as in crane girders) then it is more economic to use mild steel (grade 43) which is also convenient for very small projects or where the weight in a particular size is less than, say 5 tonnes, giving choice in obtaining material from a stockholder at short notice.

Accepted practice is to substitute a higher grade in case of non-availability of a particular steel, but in such case it is important to show the actual grade used on workshop drawings because different weld

Table 1.2 Steels to BS 4360 – summary of leading properties

Description	BS 4360 Grade	Thickness up to and including mm	Yield strength N/mm^2	Charpy test 27 joules			
						Max. thickness mm	
				Grade	Temp	Plates & sections	Hollow sections
Mild steel	43	16	275	43A**	–	–	–
		40	265	43B**	20	100	–
		63**	255	43C	0	100	40
		100**	245	43D	–20	100	40
		150† **	225	43DD	–30	100	–
				43EE	150	75	40
High tensile steel	50	16	355	50A**	–	–	–
		40	345	50B**	20	100	–
		63**	340	50C	0	100	40
		100**	325	50D	–20	100	40
		150† **	305	50DD**	–30	100	–
				50E‡ **	–40	100	–
				50EE†	–50	75	40
Weather resistant steel	WR50	12	345	50A	0	12	12
		25	345*	50B	0	50	40
		40	345*	50C	–15	50	40
		50	340§ **				

* 325 N/mm^2 in WR50A   § Up to 63 mm in WR50C. Not available in WR50A
† Not available sections   ** Not available hollow sections
‡ Not available plate

Other properties of steel:
Modulus of elasticity   E = 205 × 10^3 N/mm^2 (205 kg/mm^2)
Coefficient of thermal expansion   12 × 10^6 per °C
Density or mass   7850 kg/m^3 (7.85 tonnes/m^3 or 78.5 kN/m^3)
Elongation
(200 mm gauge length)   Grade 43   20%
                    50   18%
                    55   17%
                 WR 50   19%

procedures may be necessary. Grade 55 offers a still higher yield strength than grade 50, but it has not been widely used except for crane jibs. Table 1.3 shows typical use of steel grades and guidance is given in Table 1.4, the requirements for maximum thickness being based upon BS 5950 [2]. For bridges BS 5400 [3] has similar requirements.

Table 1.3 Main use of steel grades

	BS 4360 grade	Yield N/mm²	As rolled cost : strength ratio	Type
Buildings	43	275	1.00	Mild steel (ms)
Bridges	50	355	0.84	High tensile steels (hys)
Crane jibs	55	450	0.81	

1.2.3 Weather resistant steels

When exposed to the atmosphere, low carbon equivalent structural steels corrode by oxidation forming rust and this process will continue and eventually reduce the effective thickness leading to loss of capacity or failure. Stainless steels containing high percentages of alloying elements such as chromium and nickel can be used to virtually prevent the corrosion process but their very high cost is virtually·prohibitive for most structural purposes, except for small items such as bolts in critical locations. Protective treatment systems are generally applied to structural steel frameworks using a combination of painting, metal spraying or galvanising depending upon the environmental conditions and ease of future maintenance. Costs of maintenance can be significant for structures having difficult access conditions such as high-rise buildings with exposed frames and for bridges. Weather resistant steels which develop their own corrosion resistance and which do not require protective treatment or maintenance were developed for this reason. They were first used for the John Deare Building in Illinois in 1961, the exterior of which consists entirely of exposed steelwork and glass panels, and several prestigious buildings have used weather resistant steel frames since. The first bridge was built in 1964 in Detroit followed by many more in North America and

over 100 UK bridges have been completed since 1968. Costs of weather resistant steel frames tend to be marginally greater due to a higher material cost per tonne, but this may more than offset the alternative costs of providing protective treatment and its long term maintenance. Thus weather resistant steel deserves consideration where access for maintenance will be difficult.

Weathering resistant steels contain up to 3 per cent of alloying elements such as copper, chromium, vanadium and phosphorous. The steel oxidises naturally and when a tight patina of rust has formed this inhibits further corrosion. Figure 1.3 shows relative rates of corrosion. Over a period of one to four years the steel weathers to a shade of dark brown or purple depending upon the atmospheric conditions in the locality. Appearance is enhanced if the steel has been blast cleaned after fabrication so that weathering occurs evenly.

BS 4360 gives the specific requirements for the chemical composition and mechanical properties of the WR50 grades rolled in the UK which are similar to Corten B as originated in the USA. Mechanical properties are similar to grade 50 of BS 4360. Because the material is less widely used weathering resistant steels are not widely available from stockholders. Therefore small tonnages for a particular section or plate thickness should be avoided. Welding procedures need to be more stringent than for other high tensile steel due to the higher carbon equivalent and it must be ensured that exposed weld metal has equivalent weathering properties. Suitable alloy-bearing consumables are available for common welding processes, but for single run welds using manual or submerged arc it has been shown that sufficient dilution normally occurs such that normal electrodes are satisfactory. It is only necessary for the capping runs of butt welds to use electrodes with weathering properties.

Until the corrosion inhibiting patina has formed it should be realised that rusting takes place and run-off will occur which may cause staining of concrete and other parts locally. This can be minimised by careful attention to detail. A suitable drip detail for a bridge is shown in figure 6.8.2. Drainage of pier tops should be provided to prevent streaking of concrete and, during construction, temporary protection specified. Weather resistant steels are not suitable in conditions of total immersion or burial and therefore water traps should be avoided and columns terminated above ground level. Use of concrete or other light coloured paving should be avoided around column bases, and dark coloured brickwork or gravel finish should be considered. In the UK it is usual in bridges to design [4] against possible long term slow rusting of the steel by

Table 1.4 Guidance on steel grades on BS 5950: Part 1: 1985

Minimum structure temperature	Member		Maximum thickness mm	Steel Type	Grade	Grade – hollow sections only
0° to –5° (1)	Beams	parts in tension (5)	16 20 40 40	hys	50A 50B 50C 50D	50C 50C 50C 50D
		parts in compression	No limit	ms	50A	50C
	Members sized on deflection	parts in tension (5)	25 30 50 50 50	ms	43A 43B 43C 43D 43E	43C 43C 43C 43D 43E
		parts in compression	No limit		43A	43C
	Ties (3)		32 40 40 40	hys	50A 50B 50C 50D	50C
	Columns, struts		No limit	hys	50A	50C
	Base plates (4)		50	ms	43A	–
–15°C (2)	Beams	parts in tension (5)	10 12 27 40	hys	50A 50B 50C 50D	50C 50C 50C 50D
	Members sized on deflection	parts in compression parts in tension	No limit 15 20 40 50 50	hys ms	50A 43A 43B 43C 43D 43E	50C 43C 43C 43C 43D 43E
		parts in compression	No limit		43A	43C
	Ties (3)		20 25 40 40	hys	50A 50B 50C 50D	50C 50C 50C 50D
	Columns, struts		No limit	hys	50A	50C
	Base plates (4)		50	ms	43A	–

(1) Corresponds to internal conditions in UK

(2) Corresponds to external conditions in UK

(3) Unwelded members with any holes drilled not punched – otherwise consider thickness as for beams with stress $> 100 \ N/m^2$ parts in tension

(4) Simple baseplates not transmitting significant bending

(5) Welded members with stress $> 100 \ N/m^2$

added thicknesses (2mm for exposed face in very severe environments and 1mm otherwise), severity being a function of the atmospheric sulphur level. Weather resistant steel should not be used in marine environments and water containing chlorides such as de-icing salts should be prevented from coming into contact by suitable detailing. At expansion joints on bridges consideration should be given to casting in concrete locally in case of leakage as shown in figure 6.8.3.

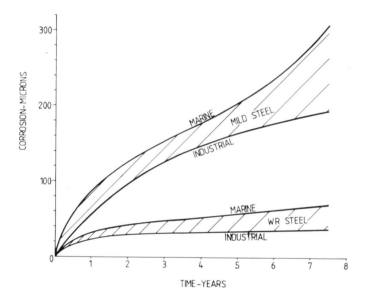

Figure 1.3 Corrosion rates of unpainted steel

Extra care must be taken in materials ordering and control during the fabrication of projects in weathering steel because its visual appearance is similar to other steels during manufacture. Testing methods are available for identification of material which may have been inadvertently misplaced.

1.3 STRUCTURAL SHAPES (see figure 1.6)

Most structures utilise hot rolled sections in the form of universal beams (UB), universal columns (UC), channels and rolled steel angles (RSA) to BS 4 [5] and BS 4848 [6]. Less frequently used are tees cut from universal

beams or columns such that the depth is one half of the original section. Hollow sections in the form of circular (CHS), square (SHS) and rectangular (RHS) shape are available but their cost per tonne is approximately 20 per cent more than universal beams and columns. Although efficient as struts or columns the end connections tend to be complex especially when bolted. They are often used where clean appearance is vital, such as steelwork which is exposed to view in public

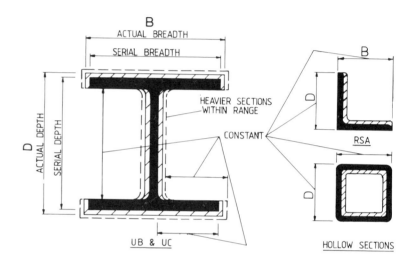

Figure 1.4 Rolled section sizes

buildings. Wind resistance is less that of open sections giving an advantage in open braced structures such as towers where the steelwork itself contributes to most of the exposed area. Other sections are available such as bulb flats as used in stiffened plate construction, for example box girder bridges and ships.

The range of universal beams (UB) and universal columns (UC) offers a number of section weights within each serial size (depth D and breadth B). Heavier sections are produced with the finishing rolls further apart such that the overall depth and breadth increases, but with the clear distance between flanges remaining constant as shown in figure 1.4. This is convenient in multi-storey buildings in allowing use of lighter sections of the same serial size for the upper levels. However, it must be remembered that the actual overall dimensions (D and B) will often be greater than the serial size except when the basic (usually lightest) section is used. This will affect detailing and overall cladding dimensions. Drawings must therefore state actual dimensions as shown in figure 6.1.6 for example. For other sections (e.g. angles and hollow sections) the overall dimensions (D and B) are constant for all weights within each serial size.

Other rolled sections are available in the UK and elsewhere including rails (for travelling cranes and railway tracks), bearing piles (H pile or welded box) and sheet piles (Larssen or Frodingham interlocking). Castellated beams are formed from universal beam or column sections cut to corrugated profile and reformed by welding to give a 50 per cent deeper section providing an efficient beam for light loading conditions.

Sections sometimes need to be curved about one or both axes to provide precamber (to counteract dead load deflection of long span beams) or to achieve permanent curvature for example in arched roofs or circular cofferdams. Specialists in the UK are able to perform section bending by a cold rolling process either to cambers (i.e. bent about the major or x–x axis) or sweeps (about the minor or y–y axis) to defined geometry achieving high accuracy of ± 2.5 mm from defined offset. The process used

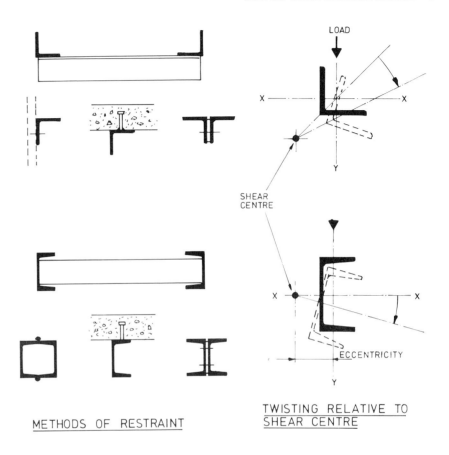

Figure 1.5 Twisting of angles and channels

Table 1.5 Sections curved about major axis. Typical minimum radii

Universal beams (UC)		Universal columns (UC)	
Designation	Minimum radius m	Designation	Minimum radius m
914 × 419 × 388	250		
914 × 305 × 289	200	356 × 368 × 202	15
838 × 292 × 226	150	305 × 305 × 283	4
762 × 267 × 197	125	254 × 254 × 167	3.5
610 × 305 × 238	40	203 × 203 × 86	3
533 × 210 × 122	25	152 × 152 × 37	2
305 × 165 × 54	10		

Note: Heaviest section in each serial size is quoted. Lighter sections may be subject to a greater minimum radius.

Information in this table is supplied by THE ANGLE RING CO. LTD, BLOOMFIELD ROAD, TIPTON, WEST MIDLANDS DY4 9EH, UK.

has merit in that most residual stresses (inherent in rolled sections when produced) are removed such that any subsequent heat inducing operations such as welding or galvanising cause less distortion than otherwise. There are practical limits to the minimum radius of curvature which can be reliably achieved, one consideration being that of local buckling. Table 1.5 indicates minimum bending radii for a number of UB or UC sections.

Fabricated members are used for spans or loads in excess of the capacity of rolled sections. Costs per tonne are higher because of the extra operations in profile cutting and welding. Box girders have particular application where their inherent torsional rigidity can be exploited, for

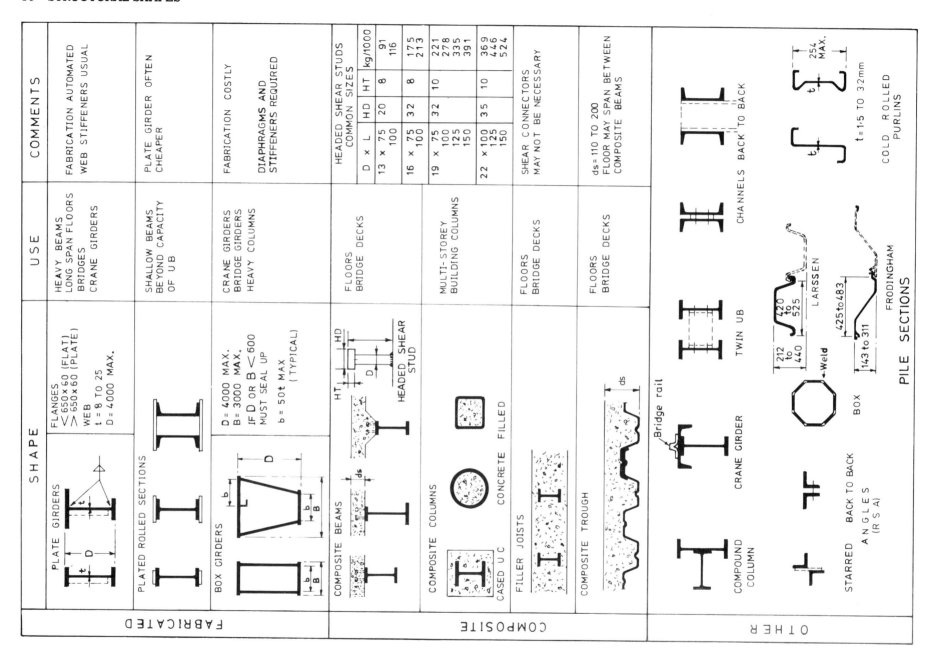

Figure 1.6 Structural shapes

HOT ROLLED SECTIONS

SHAPE	UK SIZE RANGE	USE	COMMENTS
UNIVERSAL BEAM (UB)	D × B × kg/m 127 × 76 × 13 TO 914 × 419 × 388	BEAMS	MAY NEED BEARING STIFFENERS AT SUPPORTS AND UNDER POINT LOADS B × D = SERIAL SIZE ACTUAL DIMENSIONS VARY WITH WEIGHT (kg/m)
UNIVERSAL COLUMN (UC) BEARING PILE OR H-PILE	152 × 152 × 23 TO 356 × 406 × 634	COLUMNS SHALLOW BEAMS HEAVY TRUSS MEMBERS PILES	
JOIST (8° taper)	76 × 76 × 12.65 TO 254 × 203 × 81.85	SMALL BEAMS	
CHANNEL (5° taper)	102 × 51 × 10.42 TO 432 × 102 × 65.54	BRACINGS TIES LIGHT BEAMS	WHEN USED AS BEAM TORSION OCCURS RELATIVE TO SHEAR CENTRE RESTRAIN OR USE IN PAIRS e.g.
EQUAL ANGLE (RSA) SHEAR CENTRE	D × B × t 90 × 90 × 6 TO 250 × 250 × 35	BRACINGS TOWER MEMBERS PURLINS	
UNEQUAL ANGLE (RSA)	100 × 65 × 7 TO 200 × 150 × 18	SHEETING RAILS	
STRUCTURAL TEE	D × B × kg/m UB 133 × 102 × 13 TO 419 × 457 × 194 UC 152 × 76 × 12 TO 406 × 178 × 317	TRUSS CHORDS PLATE STIFFENERS	CUT FROM UB OR UC D = 0.5 × ORIGINAL DEPTH
CASTELLATED BEAM	305 × 133 × 25 TO 1371 × 419 × 388	LIGHT BEAMS WHERE SERVICES NEED TO PASS THROUGH BEAM	MADE FROM UB $D_s = 1.5 \times D_s$ (APPROX)
BULB FLAT / PLATE	120 × 6 × 7.31 TO 430 × 20 × 90.8	PLATE STIFFENERS IN BRIDGES OR PONTOONS ETC.	GOOD WELDING ACCESS

HOLLOW SECTIONS

SHAPE	UK SIZE RANGE	USE	COMMENTS
CIRCULAR HOLLOW SECTIONS (CHS) AND TUBES	D × t CHS 21.3 × 3.2 TO 508 × 50 TUBES UP TO 2020 × 25	SPACE FRAMES COLUMNS BRACINGS PILES	END CONNECTIONS COSTLY IF BOLTED SPLICE SUPPLY COST PER TONNE APPROX 20% HIGHER THAN OPEN SECTIONS
RECTANGULAR HOLLOW SECTION (RHS)	D × B × t 50 × 25 × 2.5 TO 450 × 250 × 16	COLUMNS VIERENDEEL GIRDERS	FULLY SEAL ENDS IF EXTERNAL USE
SQUARE HOLLOW SECTION (SHS)	20 × 20 × 2 TO 400 × 400 × 16	COLUMNS SPACE FRAMES	CLEAN APPEARANCE

Figure 1.6 *Contd*

example in a sharply curved bridge. Compound members made from two
or more interconnected rolled sections can be convenient, such as twin
universal beams. For sections which are asymmetric about their major
(x–x) axis such as channels or rolled steel angles (RSA) then interconnec-
tion or torsional restraint is a necessity if used as a beam. This is to avoid
torsional instability where the shear centre of the section does not coincide
with the line of action of the applied load as shown in figure 1.5.

Cold formed sections using thin gauge material (1.5 mm to 3.2 mm
thick typically) are used for lightly loaded secondary members, such as
purlins and sheeting rails. They are not suitable for external use. They are
available from a number of manufacturers to dimensions particular to the
supplier and are usually galvanised. Ranges of standard fitments such as
sag rods, fixing cleats, cleader angles, gable posts and rafter stays are
provided, such that for a typical single storey building only the primary
members might be hot rolled sections. Detailing of cold rolled sections is
not covered in this manual, but it is important that the designer ensures
that stability is provided by these elements or if necessary provides
additional restraint.

Open braced structures such as trusses, lattice or Vierendeel girders and
towers or space frames are formed from individual members of either hot
rolled, hollow, fabricated or compound shapes. They are appropriate for
lightly loaded long span structures such as roofs or where wind resistance
must be minimised as in towers. In the past they were used for heavy
applications such as bridges, but the advent of automated fabrication
together with availability of wide plates means that plate girders are more
economic.

1.4 TOLERANCES

1.4.1 General

In all areas of engineering the designer, detailer and constructor need to
allow for tolerances. This is because in practice absolute precision cannot
be guaranteed for each and every dimension even when working to very
high manufacturing standards. Very close tolerances are demanded in
mechanical engineering applications where moving parts are involved and
the high costs involved in machining operations after manufacture of such
components have to be justified. Even here tolerance allowances are
necessary and it is common practice for values to be specified on drawings.
In structural steelwork such close tolerances could only be obtained at

very high cost, taking into account the large size of many components and
the variations normally obtained with rolled steel products. Therefore
accepted practice in the interests of economy is to fabricate steelwork to
reasonable standards obtainable in average workshop conditions and to
detail joints which can absorb small variations at site. Where justified,
operations such as machining of member ends after fabrication to precise
length and/or angularity are carried out, but this is exceptional and can
only be carried out by specialist fabricators. Normally, machining
operations should be restricted to small components (such as tapered
bearing plates) which can be carried out by a specialist machine shop
remote from the main workshop and attached before delivery to site.

Since the early 1980s many workshops have installed numerically
controlled (NC) equipment for marking, sawing members to length, for
hole drilling and profile cutting of plates to shape. This has largely
replaced the need to make wooden (or other) templates to ensure fit-up
between adjacent connections when preparation (i.e. marking cutting and
drilling) was performed by manual methods. Use of NC equipment has
significantly improved accuracy such that better tolerances are achieved
without need for adjustments by dressing or reaming of holes. However,
the main factor causing dimensional variation is *welding distortion*, which
arises due to shrinkage of the molten weld metal when cooling. The
amount of distortion which occurs is a function of the weld size, heat input
of the process, number of runs, the degree of restraint present and the
material thicknesses.

To an extent *welding distortion* can be predicted and the effects allowed
for in advance, but some fabricators prefer to exclude the use of welding
for beam/column structures and to use all bolted connections. However,
welding is necessary for fabricated sections such that the effects of
distortion must be understood and catered for.

Figure 1.7 illustrates various forms of weld distortion and how they
should be allowed for either by presetting, using temporary restraints or
initially preparing elements with extra length. This is often done at
workshop floor level, and ideally should be calculated in consultation with
the welding engineer and detailer. Where site welding is involved then the
workshop drawings should include for weld shrinkage at site by detailing
the components with extra length. A worked example is given in 1.4.2.

When site welding plate girder splices the flanges should be welded first
so that shrinkage of the joint occurs before the (normally thinner) web
joint is made to avoid buckling. Therefore the web should be detailed with

approximately 2 mm extra root gap as shown in figure 6.8.4.

Table 1.6 shows some of the main causes of dimensional variations which can occur and how they should be overcome in detailing. These practices are well accepted by designers, detailers and fabricators. It is not usual to incorporate tolerance limits on detailed drawings although this will be justified in special circumstances where accuracy is vital to connected mechanical equipment. Figure 1.7 shows tolerances for rolled sections and fabricated members.

Table 1.6 Dimensional variations and detailing practice

Type of variation	Detailing practice
1 Rolled sections – tolerances	Dimensions from top of beams down from centre of web Backmark of angles and channels
2 Length of members	Tolerance gap at ends of beams. Use lapped connections not abutting end plates For multi-storey frames with several bays consider variable tolerance packs
3 Bolted end connections	Black bolts or HSFG bolts in clearance holes For bolt groups use NC drilling or templates For large complex joints drill pilot holes and ream out to full size during a trial erection Provide large diameter holes and washer plates if excessive variation possible
4 Camber or straightness variation in members	Tolerance gap to beam splices nominal 6 mm Use lapped connections

Type of variation	Detailing practice
5 Inaccuracy in setting foundations and holding down bolts to line and level	Provide grouted space below baseplates. Cast holding down bolts in pockets. Provide extra length bolts with excess thread
6 Countersunk bolts/set screws	Avoid wherever possible
7 Weld size variation	Keep details clear in case welds are oversized
8 Columns prepared for end bearing	Machine ends of fabricated columns (end plates must be ordered extra thick) Incorporate division plate between column lengths
9 Cumulative effects on large structures	Where erection is costly or overseas delivery carry out trial erection of part or complete structure For closing piece on long structure such as bridge, fabricate or trim element to site measured dimensions
10 Fit of accurate mechanical parts to structural steelwork	Use separate bolted-on fabrication

ROLLED SECTIONS — TOLERANCES & EFFECTS IN DETAILING

CHANNELS, BEAMS & COLUMNS (BS4)

	TOLERANCES – TYPE	VALUE	DIMENSIONS
DEPTH D AT ℄ WEB	UB/UC	±3.2	FROM TOP FLANGE
	JOIST/CHANNEL D — to 305	+3.2 / −0.8	
	>305 to 406	+4.0 / −1.6	
	>406	−4.8 / −1.6	
FLANGE WIDTH B		+6.4 / −4.8	FROM ℄ OF WEB
OFF-CENTRE OF WEB e	D = >102 to 305, e = 3.2, A = D+4.8		C = [t/2 + 2mm] * N = [B−C+6mm] * n = [(D−d)/2] ‡
	D = >305, e = 4.8, A = D+6.4		ALLOW FOR TOLERANCES ASSUMING WEB TRULY VERTICAL
OUT OF SQUARENESS F+F1	B = TO 102	F+F1 = 1.6	
	>102 TO 203	3.2	
	>203 TO 305	4.8	
	>305	6.4	

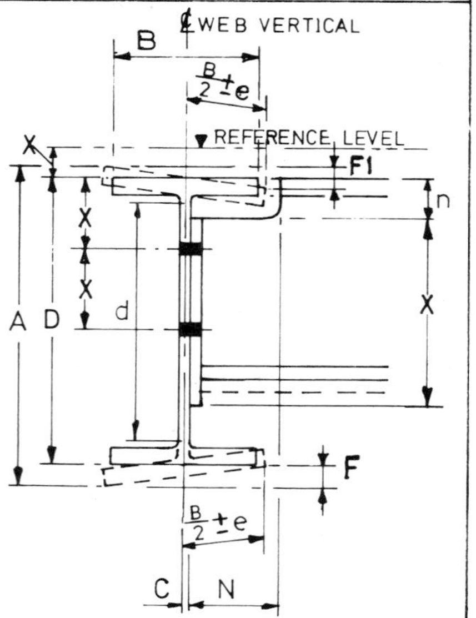

RSA CHANNELS

	TOLERANCES – TYPE	VALUE	DIMENSIONS
ROLLING TOLERANCE ON SPECIFIED WEIGHT		±2½% (BS 4)	USE BACKMARK 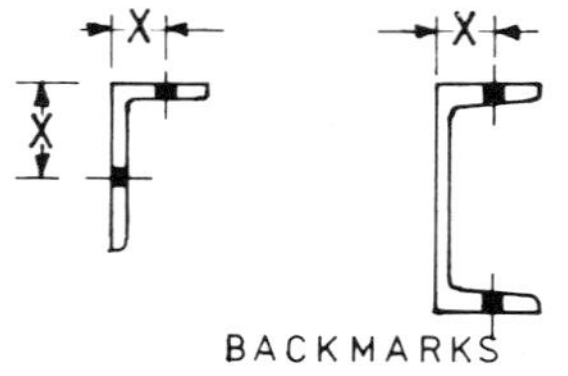

FLATS & PLATES

THICKNESS	FLATS	WIDE FLATS	PLATES	DIMENSIONS
TO 10	0.4	0.5	0.5	
>40	0.8	1.0	1.05	
>80	1.0	1.3	1.25	

WIDTH	THICK +/−		DIMENSIONS
TO 35	0.5	2% (>5mm)	
TO 150	1.5	−0, +30	OVERALL END PLATES FROM TOP FLANGE ℄ OF CROSS SECTION

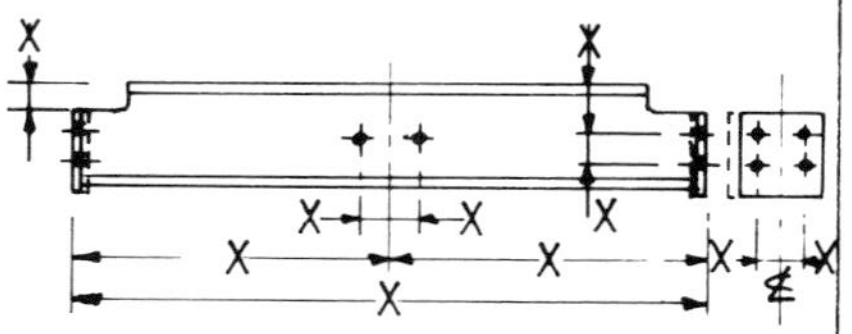

TYPICAL ONLY FROM BS4360

RANGE	TO 150	150 – 650	600 – 4000

LENGTH	+0, −3 WITH END PLATES	+4 / − WITH LAPPED END CONNECTIONS	OVERALL END PLATES

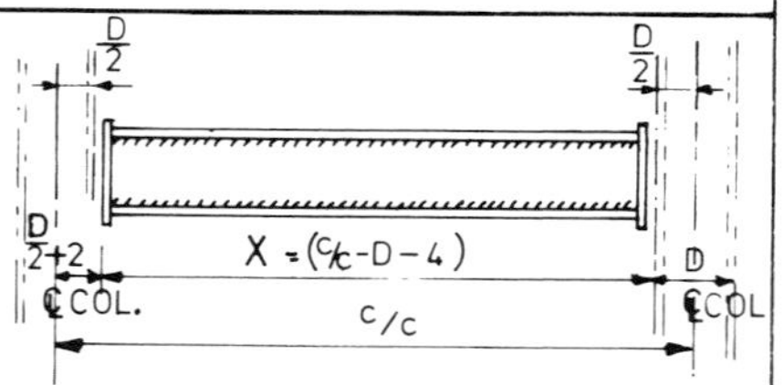

FABRICATED ITEMS

CROSS SECTION		DIMENSIONS
D	±4	FROM TOP OF TOP FLANGE
B or F	±5	FROM ℄ OF WEB
A — B to 450	+6	
A — >450	+9	
K	B/150 min.	
Δ	d/150, 3mm	

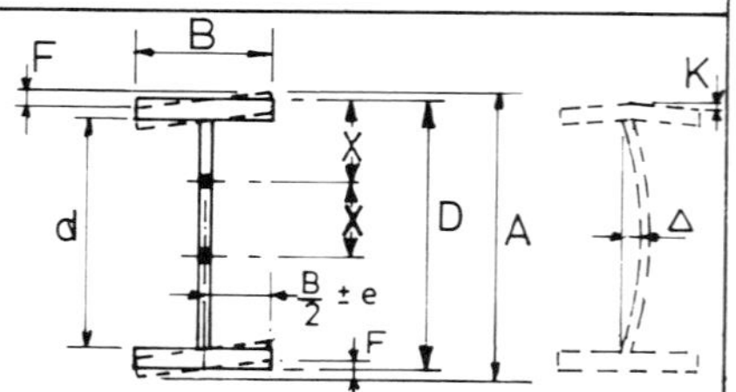

CAMBER	DEVIATION FROM CAMBER	L/1000 (MIN.12mm)	MID-LENGTH VALUES. ALSO AT STIFFENERS FOR PLATE GIRDERS. 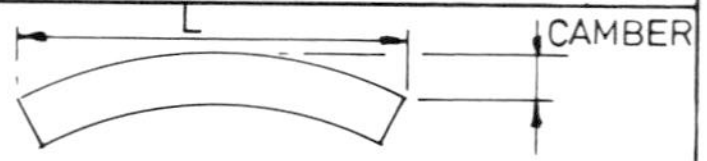
STRAIGHTNESS OR BOW		L/100 (MIN. 3mm)	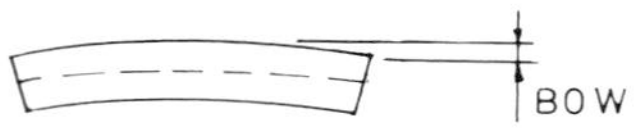

X – DIMENSIONS TO BE SHOWN ON DRAWINGS

* NEAREST 2mm ABOVE
‡ NEAREST 1mm

Figure 1.7 Tolerances

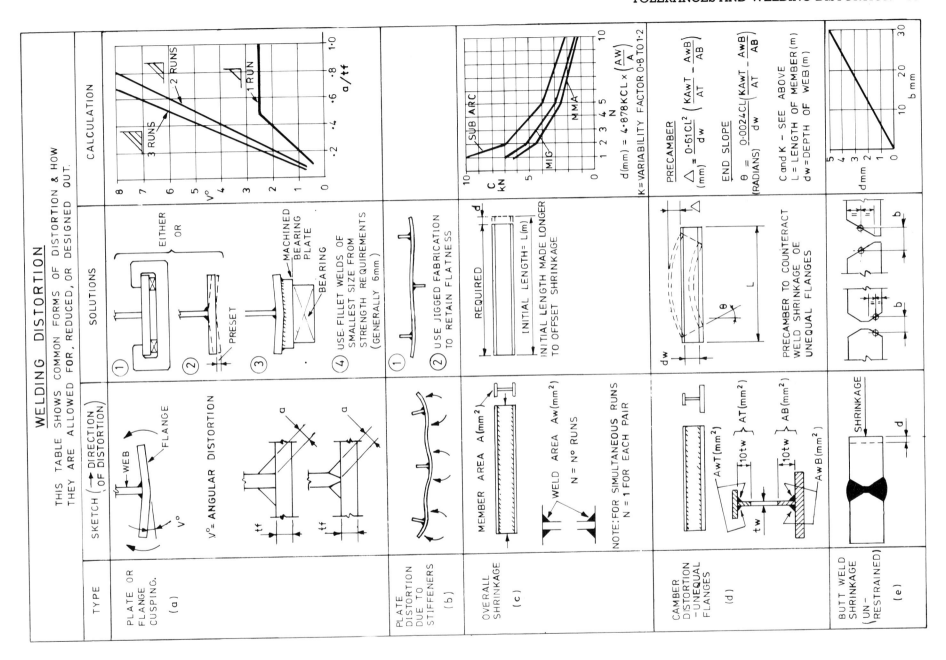

Figure 1.8 Welding distortion

1.4.2 Worked example – welding distortion for plate girder

Calculation of welding distortion
The following example illustrates use of figure 1.8 in making allowances for weld distortion for the welded plate girder shown in figure 6.8.3.

Worked example
Question
The plate girder has unequal flanges and is 32.55 m long over end plates. Web/flange welds are 8 mm fillet welds which should use the submerged arc process each completed in a single run, but not concurrently on either side of web. For simplicity the plate sizes as at mid length are assumed to apply full length. The girder as simplified is shown in figure 1.9.

Answer

(1) Amount of flange cusping
Using figure 1.8(a)
Top flange: $a = 8 \div \sqrt{2} = 5.65$ mm weld throat
$tf = 25$ mm flange thickness
$\dfrac{a}{tf} = \dfrac{5.65}{25} = 0.226$ $N = 1$ for each weld
From figure 1.8(a) $v = 1.1°$

Bottom flange: $a = 8 \div \sqrt{2} = 5.65$ mm
$tf = 50$ mm flange thickness
$\dfrac{a}{tf} = \dfrac{5.65}{50} = 0.113$ $N = 1$
From figure 1.7(a) $v = 0.5°$

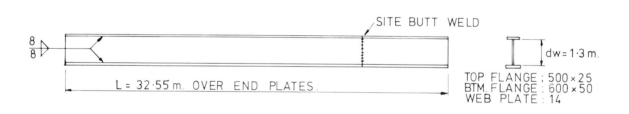

Figure 1.9 Welding distortion – worked example

It is required to calculate:

(1) Amount of flange plate cusping which may occur due to web/flange welds.
(2) Additional length of plates to counteract overall shrinkage in length due to web/flange welds.
(3) Camber distortion due to unequal flanges so that extra fabrication precamber can be determined.
(4) Butt weld shrinkage for site welded splice so that girders can be detailed with extra length.

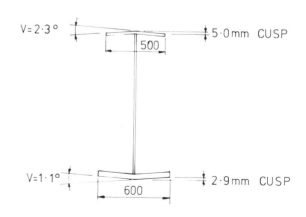

Figure 1.10 Flange cusping

Use of 'strongbacks' or presetting as shown in figure 1.8(a) may need to be considered during fabrication, because although the cusps are not detrimental structurally they may affect details especially at splices and at bearings.

(2) Overall shrinkage
Using figure 1.8(c)
Shortening $d = 4.878$ kCL (Aw/A)
where $C = 5.0$ kN for $N = 4$ weld runs

$L = 32.55$ m

$$Aw = \left(\frac{8 \times 8}{2}\right) \times 4 \text{ No} = 128 \text{ mm}^2$$

$$A = (500 \times 25) + (600 \times 50) + (1300 \times 14)$$
$$= 60\,700 \text{ mm}^2$$
$$k = 0.8 \text{ to } 1.2$$

For $k = 0.8$ $\quad d = 4.878 \times 0.8 \times 5.0 \times 32.55 \times \dfrac{128}{60700} = 1.3$ mm

or for $k = 1.2$ $\quad d = 2.0$ mm
Therefore overall length of plates must be increased by 2 mm.

(3) Camber distortion
Using figure 1.8(d)

$$\text{Precamber} = \Delta = \frac{0.61CL^2}{dw}\left(\frac{kAwt}{AT} - \frac{AwB}{AB}\right)$$

where $C = 7.0$ kN for $N = 2$ weld runs each flange

$L = 32.55$ m

$dw = 1.30$ m

$k = 0.8$ to 1.2

$$AwT = \left(\frac{8 \times 8}{2}\right) \times 2 \text{ No} = 64 \text{ mm}^2$$

$AwB = 64$ mm^2

$AT = (500 \times 25) + (10 \times 14^2) = 14\,460$ mm^2

$AB = (600 \times 50) + (10 \times 14^2) = 31\,960$ mm^2

For $k = 0.8$ $\quad \Delta = \dfrac{0.61 \times 7.0 \times 32.55^2}{1.30}\left(\dfrac{0.8 \times 64}{14\,460} - \dfrac{64}{31\,960}\right) = 5.4$ mm

For $k = 1.2$ $\quad \Delta = 11.5$ mm (say 12 mm)

End slope $\theta = \dfrac{0.0024\ CL}{dw}\left(\dfrac{kAwT}{AT} - \dfrac{AwB}{AB}\right)$

For $k = 0.8$ $\quad \theta = \dfrac{0.0024 \times 7.0 \times 32.55}{1.30}\left(\dfrac{0.8 \times 64}{14\,460} - \dfrac{64}{31\,960}\right) =$
$$0.00065 \text{ radians}$$

For $k = 1.2$ $\quad \theta = 0.00139$ radians

Therefore extra fabrication precamber needs to be applied as shown in figure 1.11 additional to the total precamber specified for counteracting dead loads etc. given in figure 6.8.1. This would not be shown on workshop drawings but would be taken account of in materials ordering and during fabrication.

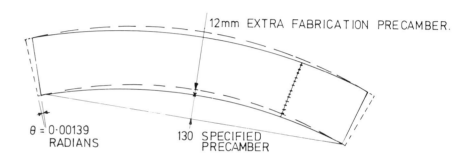

Figure 1.11 Extra fabrication precamber

(4) Butt weld shrinkage
Using figure 1.8(e)
Bottom flange. See figure 1.12 for butt weld detail.

Shrinkage $d = 2.0$ mm
Therefore length of flanges must be increased by 1 mm on each side of splice and detailed as shown. Normal practice is to weld the flanges first. Thus the web will be welded under restraint and should be detailed with the root gap increased by 2 mm as shown in figure 1.13.
In carrying out workshop drawings in this case, only item (4) should be shown thereon because items (1) to (3) occur due to fabrication effects which are allowed for at the workshop. Item (4) occurs at site and must therefore be taken into account so that the item delivered takes into account weld shrinkage at site.

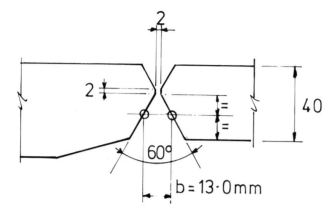

Figure 1.12 Bottom flange site weld

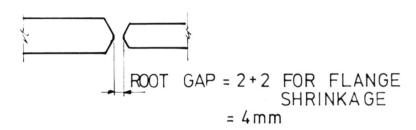

ROOT GAP = 2+2 FOR FLANGE
SHRINKAGE
= 4mm

Figure 1.13 Web site weld

1.5 CONNECTIONS

Connections are required for the functions illustrated in figure 1.14. The number of site connections should be as few as possible consistent with maximum delivery/erection sizes so that the majority of assembly is performed under workshop conditions. Welded fabrication is usual in most workshops and is always used for members such as plate girders, box girders and stiffened platework.

It is always wise to consider the connection type to be used at the conceptual design stage. A *rigidly* designed structure of lighter weight but with more complex fabrication work can be more expensive than a slightly heavier design with *simple* joints. Once the overall concept is decided the

connections should always be given at least the same attention as the design of the main members which they form. Structural adequacy is not, in itself, the sole criteria because the designer must endeavour to provide an efficient and effective structure at the lowest cost.

With appropriate stiffening either an all welded or a high strength friction grip (HSFG) bolted connection is able to achieve a fully rigid joint, that is one which is capable of developing applied bending without significant rotation. However such connections are costly to fabricate and erect. They may not always be justified. Many economical beam/column structures are built using angle cleat or welded end plate connections without stiffening and then joined with *black bolts*. These are defined as simple connections which transmit shear but where moment/rotation stiffness is not sufficient to mobilise end fixity of beams or frame action

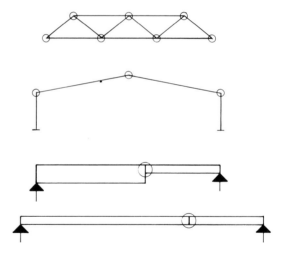

Figure 1.14 Functions of connections

under wind loading without significant deflection. Figure 1.15 shows typical moment : rotation behaviour of connections. Simple connections (i.e. types A or B) are significantly cheaper to fabricate although somewhat heavier beam sizes may be necessary because the benefits of end fixity leading to a smaller maximum bending moment are not realised. Use of simple connections enables the workshop to use automated methods more readily with greater facility for tolerance at site and will often give a more economic solution overall. However it is necessary to stabilise structures having simple connections against lateral loads such as wind by bracing or to rely on shear walls/lift cores, etc. For this reason simple connections should be made *erection-rigid* (i.e. retain resistance against free rotation whilst remaining flexible) so that the structure is stable during erection and before bracings or shear walls are connected. All connections shown in figure 1.15 are capable of being erection rigid. Calculations may be necessary in substantiation, but use of seating cleats only for beam/column connections should be avoided. A top flange cleat should be added. Web cleat or flexible (i.e. 12 mm maximum thickness) end plate connections of at least 0.6 × beam depth are suitable. Provision of seating cleats is not a theoretical necessity but they improve erection safety for high-rise structures exceeding 12 storeys. Behaviour of rigid and simple connections is shown in figure 1.16. Typical locations of site connections are shown in figure 1.17.

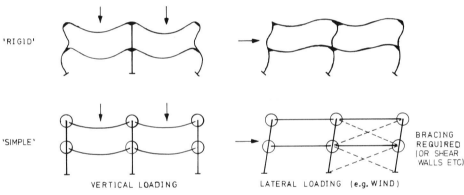

Figure 1.16 Rigid and simple connections

At site either welding or bolting is used, but the latter is faster and usually cheaper. Welding is more difficult on site because assemblies cannot be turned to permit downhand welding and erection costs arise for equipment in supporting/aligning connections, pre-heating/sheltering and non-destructive testing (NDT). The exception is a major project where such costs can be absorbed within a larger number of connections (say minimum 500). As a general rule welding and bolting are used thus:

Welding – workshop
Bolting – site

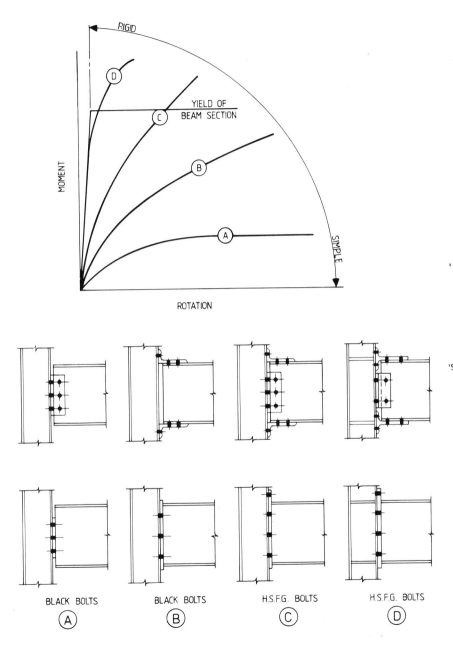

Figure 1.15 Typical moment: rotation behaviour of beam/column connections

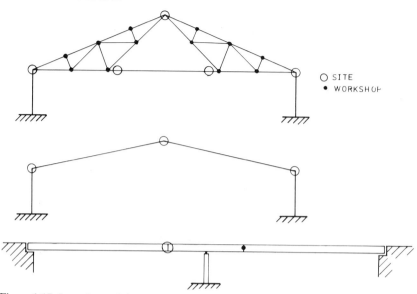

Figure 1.17 Locations of site connections

For bridges rigid connections should be used to withstand vibration from vehicular loading and spans should usually be made continuous. This allows the numbers of deck expansion joints and bearings to be reduced thus minimising maintenance of these costly items which are vulnerable to traffic and external environment.

For UK buildings connection design is usually carried out by the fabricator with the member sizes and end reactions being specified on the engineer's drawings. It is important that all design assumptions are advised to the fabricator for him to design and detail the connections. If joints are rigid then bending moments and any axial loads must be specified in addition to end reactions. For simple connections the engineer must specify how stability is to be achieved, both during construction and finally when in service.

Connections to hollow sections are generally more costly and often demand butt welding rather than fillet welds. Bolted connections in hollow sections require extended end plates or gussets and sealing plates

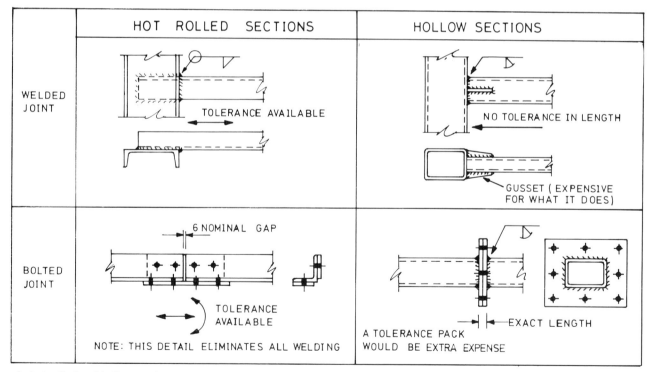

Figure 1.18 Connections in hot rolled and hollow sections

because internal access is not feasible for bolt tightening whereas channels or rolled steel angles (RSA) can be connected by simple lap joints. Figure 1.18 compares typical welded or bolted connections.

1.6 INTERFACE TO FOUNDATIONS

It is important to recognise whether the interface of steelwork to foundations must rely on a moment (or rigid) form of connection or not.

Figure 1.19 shows a steel portal frame connected either by a pin base to its concrete foundation or alternatively where the design relies on moment fixity. In the former case (a) the foundation must be designed for the vertical and horizontal reactions whereas for the latter (b) its foundation must additionally resist bending moment. In general for portal frames the steelwork will be slightly heavier with pin bases but the foundations will be cheaper and less susceptible to movements of the subsoil.

For some structures it is vital to ensure that holding down bolts are capable of providing proper anchorage arrangement to prevent uplift under critical load conditions. An example is a water tower where uplift can occur at foundation level when the tank is empty under wind loading although the main design conditions for the tower members are when the tank is full.

1.7 WELDING

1.7.1 Weld types

There are two main types of weld that is *butt weld* and *fillet weld*. A butt weld (or groove weld) is defined as one in which the metal lies substantially within the planes of the surfaces of the parts joined. It is able (if specified as a *full penetration butt weld*) to develop the strength of the parent material each side of the joint. A *partial penetration butt weld* achieves a specified depth of penetration only, where full strength of the incoming element does not need to be developed, and is regarded as a fillet weld in calculations of theoretical strength. Butt welds are shown in figure 1.20.

A fillet weld is approximately triangular in section formed within a re-entrant corner of a joint and not being a butt weld. Its strength is achieved through shear capacity of the weld metal across the throat, the weld size (usually) being specified as the leg length. Fillet welds are shown in figure 1.21.

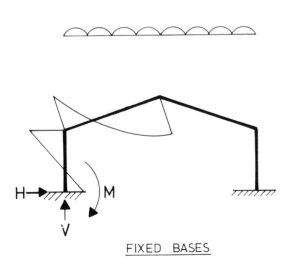

FIXED BASES

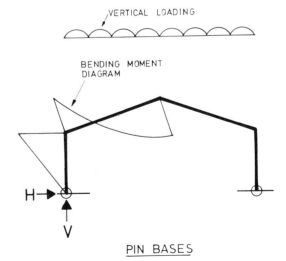

PIN BASES

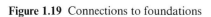

Figure 1.19 Connections to foundations

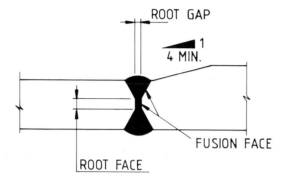

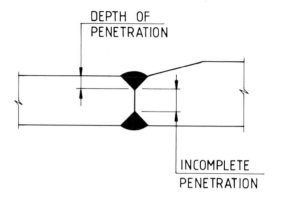

FULL PENETRATION

PARTIAL PENETRATION

Figure 1.20 Butt welds showing double v preparations

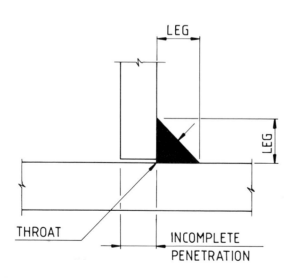

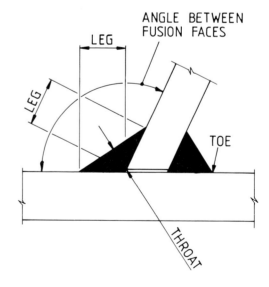

Figure 1.21 Fillet welds

1.7.2 Processes

Most workshops use electric arc manual (MMA) semi-automatic and fully automatic equipment as suited to the weld type and length of run. Either manual or semi-automatic processes are usual for short weld runs with fully automatic welding being used for longer runs where the higher rates of deposition are less being offset by extra set-up time. Detailing must allow for this. For example in fabricating a plate girder, full length web/flange runs are made first by automatic welding before stiffeners are placed with snipes to avoid the previous welding as shown in figure 1.22.

Welding processes commonly used are shown in Table 1.7.

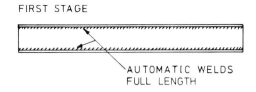

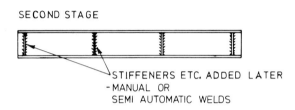

Figure 1.22 Sequence of fabrication

Table 1.7 Common weld processes

Process	Automatic or manual	Shielding	Main use	Workshop or site	Comments	Maximum size fillet weld in single run
Manual metal arc (MMA)	Manual	Flux coating on electrode	Short runs	Workshop Site	Fillet welds larger than 6 mm usually multi-run, and are uneconomic	6 mm
Submerged arc (SUBARC)	Automatic	Powder flux deposited over arc and recycled	Long runs or heavy butt welds	Workshop or site	With twin heads simultaneously welds either side of joint are possible	10 mm
Metal inert gas (MIG)	Automatic or semi-automatic	Gas (generally carbon dioxide – CO_2)	Short or long runs	Workshop	Has replaced manual welding in many workshops. Slag is minimal so galvanised items can be treated directly	8 mm

1.7.3 Weld size

In order to reduce distortion the *minimum* weld size consistent with *required* strength should be specified. The authors' experience is that engineers tend to over-design welds in the belief that they are improving the product and they often specify butt welds when a fillet weld is sufficient. The result is a more expensive product which will be prone to unwanted distortion during manufacture. This can actually be detrimental if undesirable rectification measures are performed especially at site, or result in maintenance problems due to lack of fit at connections. An analogy exists in the art of the dressmaker who sensibly uses fine sewing thread to join seams to the thin fabric. She would never use strong twine, far stronger, but which would tear out the edges of the fabric, apart from being unsightly and totally unnecessary.

Multiple weld-runs are significantly more costly than single run fillet welds and therefore joint design should aim for a 5 mm or 6 mm leg except for long runs which will clearly be automatically welded when an 8 mm or 10 mm size may be optimum depending upon design requirements. For light fabrication using hollow sections with thickness 4 mm or less, then 4 mm size should be used where possible to reduce distortion and avoid burn-through. For thin platework (8 mm or less) the maximum weld size should be 4 mm and use of intermittent welds (if permitted) helps to reduce distortion. If it is to be hot-dip galvanised then distortion due to release of residual weld stresses can be serious if large welds are used with thin material. Intermittent welds should not be specified in exposed situations (because of corrosion risk) or for joints which are subject to fatigue loading such as crane girders, but are appropriate for internal areas of box girders and pontoons.

1.7.4 Choice of weld type

Butt welds, especially full penetration butt welds, should only be used where essential such as in making up lengths of beam or girder flange into full strength members. Their high cost is due to the number of operations necessary including edge preparation, back gouging, turning over, grinding flush (where specified) and testing, whereas visual inspection is often sufficient for fillet welds. Welding of end plates, gussets, stiffeners, bracings and web/flange joints should use fillet welds even if more material is implied. For example lapped joints should always be used in preference to direct butting as shown in figure 1.23.

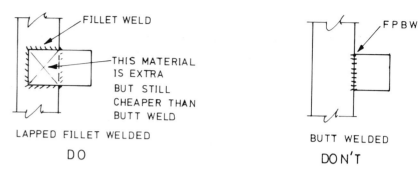

Figure 1.23 Welding using lapped joints

In the UK welding of structural steel is carried out to BS 5135:1984 [7] which requires weld procedures to be drawn up by the fabricator. It includes recommendations for any preheating of joints so as to avoid hyydrogen induced cracking, this being sometimes necessary for high tensile steels. Fillet welds should where possible be returned around corners for a length of at least twice the weld size to reduce the possibility of failure emanating from weld terminations, which tend to be prone to start : stop defects.

1.7.5 Lamellar tearing

In design and detailing it should be appreciated that structural steels, being produced by rolling, possess different and sometimes inferior mechanical properties transverse to the rolled direction. This occurs because non-metallic manganese sulphides and manganese silica inclusions which occur in steel making become extended into thin planar type elements after rolling. In this respect the structure of rolled steel resembles timber to some extent in possessing grain direction. In general this is not of great significance from a strength viewpoint. However, when large welds are made such that a fusion boundary runs parallel to the planar inclusion, the phenomenon of *lamellar tearing* can result. Such tearing is initiated and propagated by the considerable contractile stress across the thickness of the plate generated by the weld on cooling. If the joint is under *restraint* when welded, such as when a cruciform detail is welded which is already assembled as part of a larger fabrication then the possibility of lamellar tearing cannot be ignored. This is exacerbated where full penetration butt welds are specified not only because of the greater volume of weld metal involved, but because further transverse strains will be caused by the heat input of back-gouging processes used between weld runs to ensure fusion. The best solution is to avoid cruciform welds having full penetration butt welds. If cruciform joints are unavoidable then the thicker of the two plates should pass through, so that the strains which occur during welding are less severe. In other cases a special through thickness steel grade can be specified which has been checked for the presence of lamination type defects. However, the ultrasonic testing which is used may not always give

a reliable guide to the susceptibility to lamellar tearing. Fortunately most known examples have occurred during welding and have been repaired without loss of safety to the structure in service. However, repairs can be extremely costly and cause unforeseen delays. Therefore details which avoid the possibility of lamellar tearing should be used whenever possible. Figure 1.24 shows lamellar tearing together with suggested alternative details.

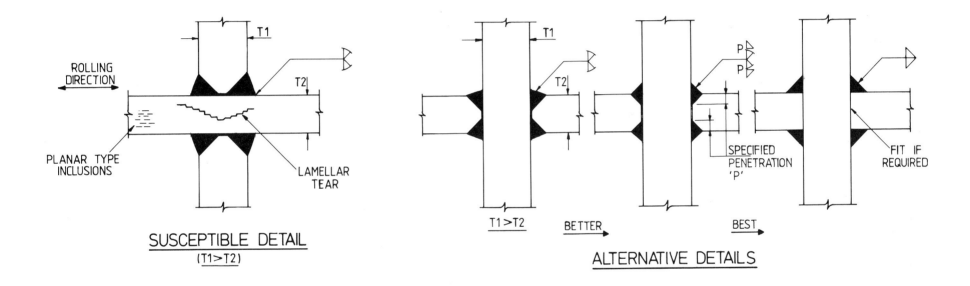

Figure 1.24 Lamellar tearing

1.8 BOLTING

1.8.1 General

Bolting is the usual method for forming site connections and is sometimes used in the workshop. The term 'bolt' used in its generic sense means the assembly of bolt, nut and appropriate washer. Bolts in clearance holes should be used except where absolute precision is necessary. *Black bolts* (the term for an untensioned bolt in a clearance hole 2 or 3 mm larger than the bolt dependent upon diameter) can generally be used except in the following situations where slip is not permissible at working loads:

(1) Rigid connections – for bolts in shear.
(2) Impact, vibration and fatigue-prone structures – e.g. silos, towers, bridges.
(3) Connections subject to stress reversal (except where due to wind loading only).

High strength friction grip (HSFG) bolts should be used in these cases or, exceptionally, precision bolts in close tolerance holes (+0.15 mm – 0 mm) may be appropriate.

If bolts of different grade or type are to be used on the same project then it is wise to use different diameters. This will overcome any possible errors at the erection stage and prevent incorrect grades of bolt being used in the holes. For example a typical arrangement would be:

All grade 4.6 bolts – 20 mm diameter
All grade 8.8 bolts – 24 mm diameter

Black bolts and HSFG bolts are illustrated in figure 1.25. The main bolt types available for use in the UK are shown in Table 1.8.

The European continent system of strength grading introduced with the ISO system is given by two figures, the first being one tenth of the minimum ultimate stress in kgf/mm^2 and the second is one tenth of the percentage of the ratio of minimum yield stress to minimum ultimate. Thus '4.6 grade' means that the minimum ultimate stress is 40 kgf/mm^2 and the yield stress 60 per cent of this. The yield stress is obtained by multiplying the two figures together to give 24 kgf/mm^2. For higher tensile products where the yield point is not clearly defined, the stress at a permanent set limit is quoted instead of yield stress.

Table 1.8 Bolts used in UK

Type	BS No	Main use	Workshop or site
Black bolts, grade 4.6 (mild steel)	BS 4190 [8] (nuts and bolts) BS 4320 [9] (washers)	As black bolts in clearance holes	Workshop or site
High tensile bolts, grade 8.8	BS 3692 [10] (nuts and bolts) BS 4320 (washers)	As black bolts in clearance holes As precision bolts in close tolerance holes	Workshop or site Workshop
HSFG bolts, general grade	BS 4395 [11] Pt 1 (bolts, nuts and washers)	Bolts in clearance holes where slip not permitted. Used to BS 4604 [12] Pt 1	Workshop or site
Higher grade	BS 4395 [11] Pt 2 (bolts, nuts and washers)	Bolts in clearance holes where slip not permitted. Used to BS 4604 [12] Pt 2	Workshop or site
Waisted shank	BS 4395 [11] Pt 3 (bolts, nuts and washers)	Bolts in clearance holes where slip not permitted. Used to BS 4604 [12] Pt 3	Little used

BLACK BOLTS

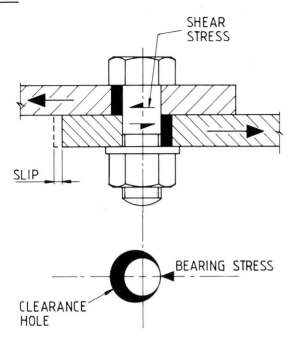

H.S.F.G. BOLTS

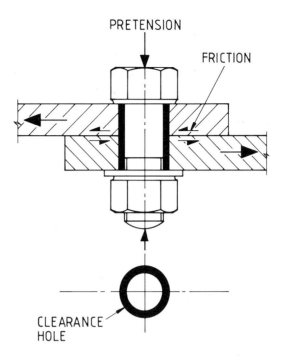

Figure 1.25 Black bolts and HSFG bolts

The single grade number given for nuts indicates one tenth of the proof load stress in kgf/mm^2 and corresponds with the bolt ultimate strength to which it is matched, e.g. an 8 grade nut is used with an 8.8 grade bolt. It is permissible to use a higher strength grade nut than the matching bolt number and grade 10.9 bolts are supplied with grade 12 nuts since grade 10 does not appear in the British Standard series. To minimise risk of thread stripping at high loads, BS 4395 high strength friction grip bolts are matched with nuts one class higher than the bolt.

1.8.2 High strength friction grip (HSFG) bolts

A pre-stress of approximately 70 per cent of ultimate load is induced in the shank of the bolts to bring the adjoining plies into intimate contact. This enables shear loads to be transferred by friction between the interfaces and makes for rigid connections resistant to movement and fatigue. HSFG bolts thus possess the attributes possessed by rivets, which welding and

bolts displaced during the early 1950s.

During tightening the bolt is subjected to two force components:

(1) The induced axial tension.
(2) Part of the torsional force from the wrench applied to the bolt via the nut thread.

The stress compounded from these two forces is at its maximum when tightening is being completed. Removal of the wrench will reduce the torque component stress, and the elastic recovery of the parts causes an immediate reduction in axial tension of some 5 per cent followed by further relaxation of about 5 per cent, most of which takes place within a few hours. For practical purposes, this loss is of no consequence since it is taken into account in the determination of the slip factor, but it illustrates that a bolt is tested to a stress above that which it will experience in service. It may be said that if a friction grip bolt does not break in tightening, the likelihood of subsequent failure is remote. The bolt remains in a state of virtually constant tension throughout its working life. This is most useful for structures subject to vibration, e.g. bridges and towers. It also ensures that nuts do not become loose with risk of bolt loss during the life of the structure, thus reducing the need for continual inspection.

Mechanical properties for general grade HSFG bolts (to BS 4395: Part 1 [11]) are similar to grade 8.8 bolts for sizes up to and including M24. Although not normally recommended, grade 8.8 bolts can exceptionally be used as HSFG bolts.

HSFG bolts may be tightened by three methods, viz:

(1)   Torque control
(2)   Part turn method
(3)   Direct tension indication.

The latter is now usual practice in the UK and the well-established 'Coronet'* load indicator is often used which is a special washer with arched protrusions raised on one face. It is normally fitted under the standard bolt head with the protrusions facing the head thus forming a gap between the head and load indicator face. On tightening the gap reduces as the protrusions depress and when the specified gap (usually 0.40 mm) is obtained, the bolt tension will not be less than the required minimum. Assembly is shown in figure 1.26.

* 'Coronet' load indicators are manufactured by Cooper & Turner Limited, Vulcan Works, Vulcan Road, Sheffield S9 2FW, United Kingdom.

1.9 DOS AND DON'TS

The overall costs of structural steelwork are made up of a number of elements which may vary considerably in proportion depending upon the type of structure and site location. However a typical split is shown in Table 1.9.

It may be seen that the materials element (comprising rolled steel from the mills, bolts, welding consumables, paint and so on) is significant, but constitutes considerably less in proportion than the workmanship. This is why the economy of steel structures depends to a great extent on *details* which allow easy (and therefore less costly) fabrication and erection. Minimum material content is important in that designs should be efficient,

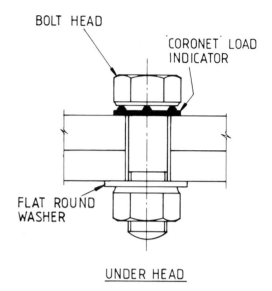

UNDER HEAD

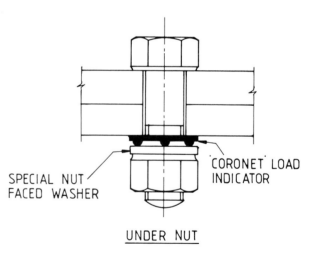

UNDER NUT

Figure 1.26 Use of 'Coronet' load indicator

Table 1.9 Typical cost proportion of steel structures

	Materials %	Workmanship %	Total %
Materials	30	0	30
Fabrication	0	45	45
Erection	0	15	15
Protective treatment	5	5	10
Total	35	65	100

but more relevant is the correct selection of structural type and fabrication details. The use of automated fabrication methods has enabled economies to be made in overall costs of steelwork, but this can only be realised fully if details are used which permit tolerance (see section 1.4) so that time consuming (and therefore costly) rectification procedures are avoided at site. Often if site completion is delayed then severe penalties are imposed on the steel contractor and this affects the economy of steelwork in the long term.

For this reason one of the purposes of this manual is to promote the use of details which will avoid problems both during fabrication and erection. Figures 1.27 and 1.28 show a series of dos and don'ts which are intended to be used as a general guide in avoiding uneconomic details. Figure 1.29 gives dos and don'ts related to corrosion largely so as to permit maintenance and avoid moisture traps.

1.10 PROTECTIVE TREATMENT

When exposed to the atmosphere all construction materials deteriorate with time. Steel is affected by atmospheric corrosion and normally requires a degree of protection, which is no problem but requires careful assessment depending upon:

Aggressiveness of environment
Required life of structure
Maintenance schedule
Method of fabrication and erection
Aesthetics.

It should be remembered that for corrosion to occur air and moisture both need to be present. Thus, permanently embedded steel piles do not corrode, even though in contact with water, provided air is excluded by virtue of impermeability of the soil. Similarly the internal surfaces of hollow sections do not corrode provided complete sealing is achieved to prevent continuing entry of moist air.

There is a wide selection of protective systems available, and that used should adequately protect the steel at the most economic cost. Detailing has an important influence on the life of protective treatment. In particular details should avoid the entrapment of moisture and dirt between profiles or elements especially for external structures. Figure 1.29 gives some dos and don'ts related to corrosion. Provided that the ends are sealed by welding, then hollow sections do not require treatment internally. For large internally stiffened hollow members which contain internal stiffening such as *box girder bridges* and *pontoons* needing future inspection, it is usual to provide an internal protective treatment system. Access manholes should be sealed by covers with gaskets to prevent ingress of moisture as far as possible, allowing use of a cheaper system. For immersed structures such as pontoons which are inaccessible for maintenance, corrosion prevention by cathodic protection may be appropriate.

Adequate preparation of the steel surface is of the utmost importance before application of any protective system. Modern fabricators are properly equipped in this respect such that the life of systems has considerably extended. For external environments it is especially essential that all millscale is removed which forms when the hot surface of rolled steel reacts with air to form an oxide. If not removed it will eventually become detached through corrosion. Blast cleaning is widely used to prepare surfaces, and other processes such as hand cleaning or flame cleaning are less effective although acceptable in mild environments. Various national standards for the quality of surface finish achieved by blast cleaning are correlated in Table 1.10.

Table 1.10 National standards for grit blasting

British Standard	Swedish Standard	USA Steel Structures Painting Council
BS 4232 [13]	SIS 05 59 00 [14]	SSPC [15]
1st quality	Sa 3	White metal
2nd quality	Sa 2½	Near white
3rd quality	Sa 2	Commercial

A brief description is given for a number of accepted systems in Table 1.11 based on UK conditions to BS 5493 [16] and Department of Transport guidance [17]. Specialist advice may need to be sought in particular environments or areas.

The following points should be noted when specifying systems:

(1) Metal coatings such as galvanising and aluminium spray give a durable coating more resistant to site handling and abrasion but are generally more costly.

(2) Galvanising is not suitable for plate thicknesses less than 5 mm. Welded members, especially if slender, are liable to distortion due to release of residual stress and may need to be straightened. Galvanising is especially suitable for piece-small fabrications which may be vulnerable to handling damage, such as when despatched overseas. Examples are towers or lattice girders with bolted site connections. The maximum size is dependent upon the available size of galvanising baths. Indicative UK maximum sizes of assemblies are:

$$16.8 \text{ m} \times 3.6 \text{ m} \times 1.5 \text{ m}$$
$$\text{or } 5.2 \text{ m} \times 3.6 \text{ m} \times 1.8 \text{ m}.$$

(3) For HSFG bolted joints the interfaces must be grit blasted 2nd quality or metal sprayed only, without any paint treatment to achieve friction. A reduced slip factor must be assumed for galvanised steelwork. During painting in the workshop the interfaces are usually masked with tape which is removed at site assembly. Paint coats are normally stepped back at 30 mm intervals, with the first coat taken 10 to 15 mm inside the joint perimeter. Sketches may need to be prepared to define painted/masked areas.

(4) For non friction bolted joints the first two workshop coats should be applied to the interfaces.

(5) Micaceous iron oxide paints are obtainable in limited colour range only (e.g. medium grey, natural grey, dark blue, dark green) and provide a matt finish. Where a decorative or gloss finish is required then another system of overcoating must be used.

(6) Surfaces in contact with concrete should be free of loose scale and rust but may otherwise be untreated. Treatment on adjacent areas should be returned for at least 25 mm and any metal spray coating must be overcoated.

(7) Treatment of bolts at site implies blast cleaning unless they have been galvanised. As an alternative, consideration can be given to use of electro-plated bolts, degreased after tightening followed by etch priming and painting as for the adjacent surfaces.

(8) Any delay between surface preparation and application of the first treatment coat must not normally exceed four hours.

(9) Lifting cleats should be provided for large fabrications exceeding say 10 tonnes in weight to avoid handling damage.

(10) The maximum amount of protective treatment should be applied at workshop in enclosed conditions. Generally it is convenient to apply at least the final paint coat at site after making good any erection damage.

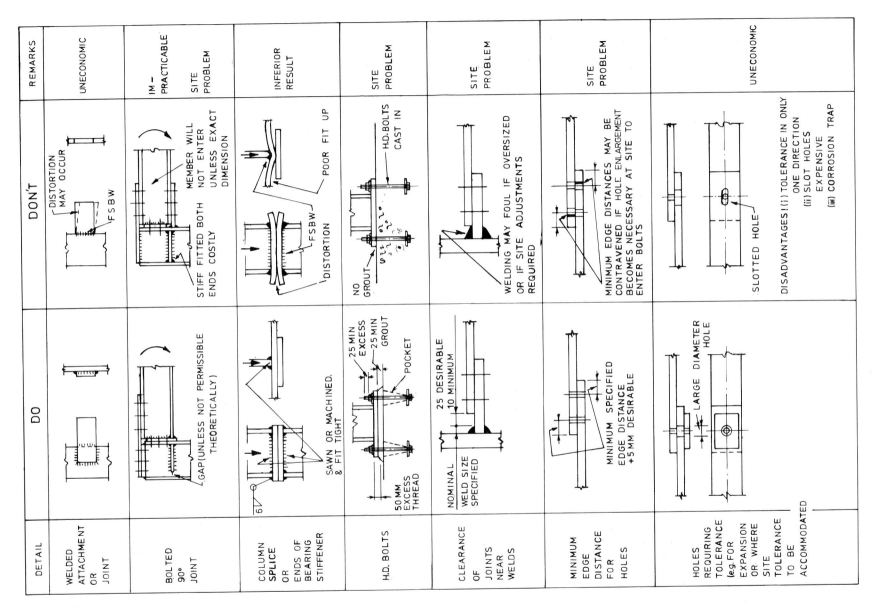

Figure 1.27 Dos and don'ts

32 DOS AND DON'TS

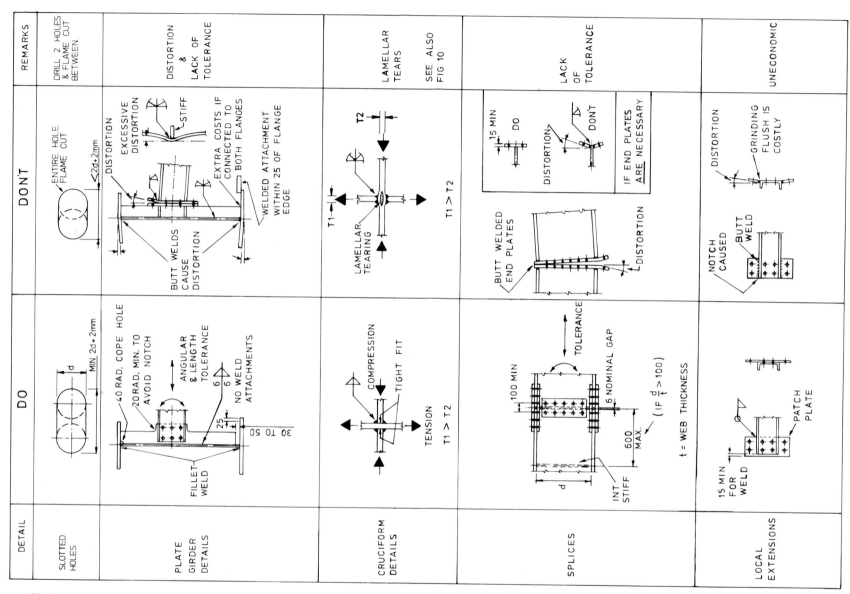

Figure 1.28 Dos and don'ts

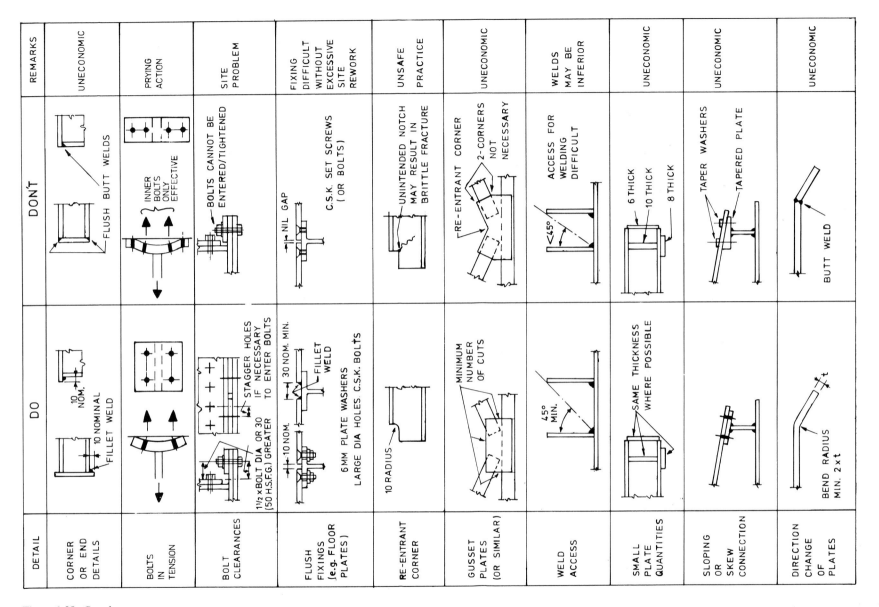

Figure 1.28 *Contd*

34   DOS AND DON'TS

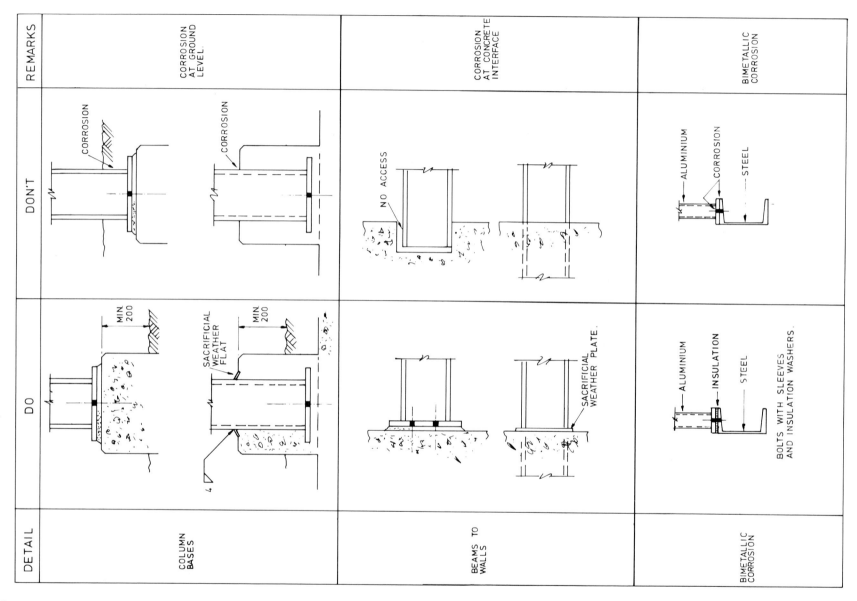

Figure 1.29 Dos and don'ts – corrosion

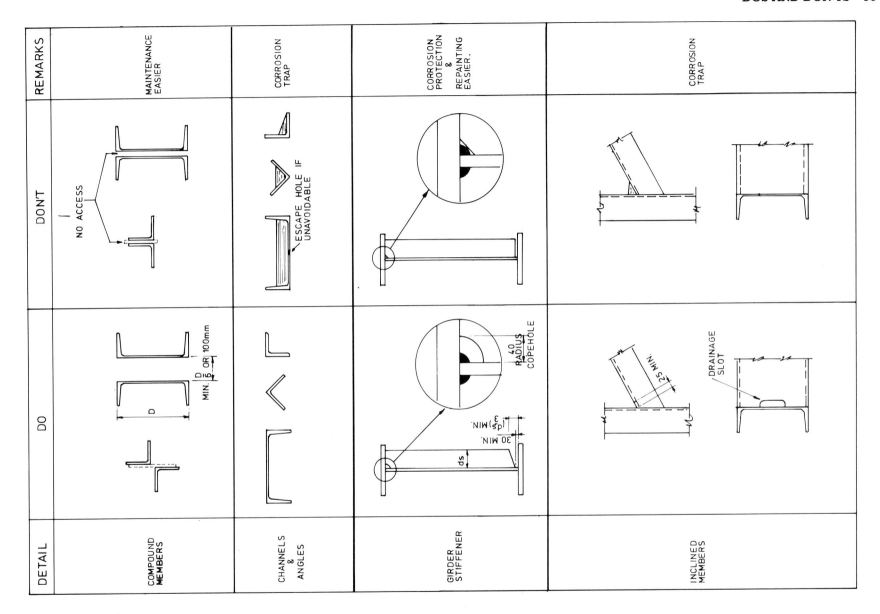

Figure 1.29 *Contd*

Table 1.11 Typical protective treatment systems

Environment (i)	Time (ii)	Structure type	Surface preparation & minimum quality (iii)	Metal coating & minimum thickness (microns)	Paint coats (iv) 1	2	3	4	5	6	Total paint dry film thickness (microns)	Designation BS 5493 or DTp (v)	Treatment of bolts (vi)
DI	10–20	Buildings	2nd	–	zre pp	zre	–	–	–	–	75 nominal	SD2 / –	Paint
WI	10–20	Buildings	2nd	–	zre pp	zre	zre	–	–	–	150 nominal	SD5 / –	Paint
SZ	10–20	Marine structures	2nd	Aluminium spray 100	aes	PE	PE	–	–	–	225 min	SC10A / –	Galvanise Paint (vii)
SI	10–20	Marine structures	2nd	–	zp	PE	PE	AF	–	–	310 min	SK8 / –	Galvanise Paint (vii)
NI	12	General	2nd	–	zpe pp	HB zpa	HB/a (S)	–	–	–	140 nominal	SF6* / –	Paint
NI NC	> 20	Piece-small fabrication	–	Galvanise 85	–	–	–	–	–	–	–	SB1* / –	Galvanise
NI NC PI PC	15	Piece-small fabrication	–	Galvanise 85	T-wash (S)	HB CR (S)	–	–	–	–	75 nominal	SB9* / –	Galvanise Paint
NI NC PI PC	15	General	2nd	–	zre pp	zre	MI0/e	HB CR (S)	–	–	290 nominal	–* / –	Paint
NI NC PI PC	15	Piece-small fabrication	3rd	Galvanise 140	–	–	–	–	–	–	–	SB2* / –	Galvanise
NI NC PI PC	18	General	1st	Aluminium spray 150	Sealer	–	–	–	–	–	–	SC6* / – / Type 1	Plate Aluminium spray
NI NC PI PC	20	General	1st	–	ezg pp	ezg	CR/a	HB MI0 CR (S)	HB MI0 CR (S)	–	–	–* / –*	Galvanise / Paint (vii)
NI	12	Bridges for Department of Transport	2nd	–	RO/ZC	zpe	zpe	MI0/p	MI0/p S	–	200 min	– / Type 2	Paint
NI PI	12		2nd	–	zp CR/a	zp CR/a	zp CR/a	MI0 CR S	CR S	–	250 min	– / Type 3	Paint
NI PI	12		2nd	–	zp AR	zp AR	zp AR	MI0 AR S	AR	–	250 min	– / Type 4	Paint
NI PI	12		1st	Aluminium spray 100	aes	zp CR/a	MI0 CR S	CR S	–	–	200	Type 5	Aluminium spray & Paint

Table 1.11 *Contd*

Environment (i)	Time (ii)	Structure type	Surface preparation & minimum quality (iii)	Metal coating & minimum thickness (microns)	Paint coats (iv)						Total paint dry film thickness (microns)	Designation BS 5493 or DTp (v)	Treatment of bolts (vi)
					1	2	3	4	5	6			
NI PI	12		1st	Aluminium spray 100	aes	zp AR	MI0 AR S	AR S	–	–	200	Type 6	Aluminium spray & Paint
NI NC PI PC	12		2nd	–	zp CR/a	zp CR/a	zp CR/a	MI0 CR	CR S	CR S	300	Type 7	Paint
NI NC PI PC	12		2nd	–	zP AR	zp AR	zp AR	MI0 AR	AR S	AR S	300	Type 8	Paint
NI NC PI PC	12	Bridges for Department of Transport	1st	Aluminium spray 100	aes	zp CR/a	zp CR/a	MI0 CR S	CR S	–	250	Type 9	Aluminium spray & Paint
NI NC PI PC	12		1st	Aluminium spray 100	aes	zp AR	zp AR	MI0 AR S	AR S	–	250	Type 10	Aluminium spray & Paint
Interior of box girders	12		2nd	–	RO/ZC	zpe	MI0/p	MI0/p S	–	..	175	Type 11	Paint
NI RS	12	Bridge parapets	2nd	Galvanise 85	T-wash	zpc (S)	MI0/p (S)	MI0/p (S)	–	–	175	Type 12	Galvanise & Paint
NI NC RS PI PC	12	Bridge parapets	2nd	Galvanise 85	T-wash	zp AR	MI0 AR	MI0 AR (S)	–	–	200	Type 13	Galvanise & Paint

Notes to Table 1.11

(i) Environments:

 DI Dry internal
 WI Wet or damp internal
 NI Normal inland (most rural and urban areas)
 NC Normal coastal (up to 3 km inland)
 PI Polluted inland (high airborne sulphur dioxide or other pollution)
 PC Polluted coastal
 RS Road (de-icing) salts
 SI Sea water immersed
 SS Sea water splash zone

(ii) Time indicated is approximate period in years to first major maintenance and is appropriate to the most severe environment indicated. The time will be subject to variation depending upon the micro-climate around the structure. Maintenance may need to be more frequent for decorative appearance.

(iii) Surface preparation quality refers to BS 4232 [13].

(iv) The number of coats given is indicative. A different number of coats may be necessary depending upon the method of application in order to comply with the dry film thickness specified.

(v) Based on UK Department of Transport Notes for Guidance [17].

(vi) Paint treatment for exposed parts of bolts, nuts and washers to be as for structure, except that prefabrication primer may be omitted. Department of Transport also require two extra stripe coats at all fasteners, welds and external corners.

(vii) Paint treatment over galvanised bolts preceeded by T-Wash or polyvinyl butyral etch primer.

Abbreviations for paint coats

zp	zinc phosphate
zpa	zinc phosphate modifier alkyd
zpe	zinc phosphate epoxy ester
zre	zinc rich epoxy (2 pack)
MI0/e	epoxy micaceous iron oxide (2 pack)
MI0/p	micaceous iron oxide phenolic
ezs	ethyl zinc silicate
HB	high build
S	applied at site
(S)	applied at site optionally
CR	chlorinated rubber
AR	acrylated rubber
pp	prefabrication primer (only required where surface preparation precedes fabrication)
PE	pitch epoxy
RO/ZC	red oxide/zinc chromate primer
AF	anti fouling
T-Wash	A non-proprietary material known as British Rail T-Wash which turns zinc surfaces black when properly applied
aes	aluminium epoxy sealer (2 pack)
/a	alkyd

1.11 DRAWINGS

1.11.1 Engineer's drawings

Engineer's drawings are defined as the drawings which describe the employer's requirements and main details. Usually they give all leading dimensions of the structure including alignments, levels, clearances, member size and show steelwork *in an assembled form*. Sometimes, especially for buildings, connections are not indicated and must be designed by the fabricator to forces shown on the engineer's drawings requiring submission of calculations to the engineer for approval. For major structures such as bridges the engineer's drawings usually give details of connections including sizes of all bolts and welds. Most example drawings of typical structures included in this manual can be defined as engineer's drawings.

Engineer's drawings achieve the following purposes:

(1) Basis of engineer's cost estimate before tenders are invited.
(2) To invite tenders upon which competing contractors base their prices.
(3) Instructions to the contractor during the contract (i.e. *contract drawings*) including any revisions and variations. Most contracts usually involve revisions at some stage due to employer's amended requirements or due to unexpected circumstances such as variable ground conditions.
(4) Basis of measurement of completed work for making progressive payments to the contractor.

1.11.2 Workshop drawings

Workshop drawings (or shop details) are defined as the drawings prepared by the contractor (i.e. the fabricator, often in capacity of a subcontractor) showing each and every component or member in full details for fabrication. A requirement of most contracts is that workshop drawings are submitted to the engineer for approval, but that the contractor remains responsible for any errors or omissions. Most responsible engineers nevertheless carry out a detailed check of the workshop drawings and point out any apparent shortcomings. In this way any undesirable details are hopefully discovered before fabrication and the chance of error is reduced. Usually a marked copy is returned to the contractor who then amends the drawings as appropriate for re-submission. Once approved the workshop drawings should be correctly regarded as contract drawings.

Workshop drawings are necessary so that the contractor can organise efficient production of large numbers of similar members, but with each having slightly different details and dimensions. Usually each member is shown fabricated as it will be delivered on site. Confusion and errors can be caused under production conditions if only typical drawings showing many variations, lengths and 'opposite handing' for different members are issued. *Workshop drawings of members must include reference dimensions to main grid lines* to facilitate cross referencing and checking. This is difficult to undertake without the possibility of errors if members are drawn only in isolation. All extra welds or joints necessary to make up member lengths must be included on workshop drawings. Marking plans must form part of a set of workshop drawings to ensure correct assembly and to assist planning for production, site delivery and erection. A General Arrangement drawing is often also required giving overall setting out including holding down bolt locations from which workshop drawing lengths, skews and connections have been derived. Often the engineer's drawings are inadequate for this purpose because only salient details and overall geometry will have been defined.

Workshop drawings must detail camber geometry for girders so as to counteract (where required and justified) dead load deflection, including the correct inclinations of bearing stiffeners. For site welded connections the workshop drawings must include all temporary welding restraints for attachment and joint root gap dimensions allowing for predicted weld shrinkage. Each member must be allocated a mark number. A system of 'material marks' is also usual and added to the workshop drawings so that each stiffener or plate can be identified and cut by the workshop from a material list.

1.12 CODES OF PRACTICE

In the UK appropriate Codes of Practice for design and construction of steelwork are as summarised below and these Codes are often suitable in other countries. Different design criteria may need to be applied for example in the cases of varied loadings, earthquake effects, temperature range and so on.

1.12.1 Buildings

Steelwork in buildings is designed and constructed in the UK to BS 5950. Part 1 [2] published in 1985 is a Code for the design of hot rolled sections in buildings. A guide is available [18] giving member design capacities, together with those for bolts and welds. BS 5950 Part 2 [2] is a specification for materials fabrication and erection, and BS 5493 [15] gives guidance on protective treatment. BS 5950 Part 5 [2] deals with cold formed sections. In the interim period until publication of the part dealing with composite construction CP 117 Part 1 [19] and Constrado recommendations [20] are available.

BS 5950 uses the *limit state* concept in which various limiting states are considered under factored loads. The main limit states are:

Ultimate limit state	*Serviceability limit state*
Strength (i.e. collapse)	Deflection
Stability (i.e. overturning)	Vibration
Fatigue fracture	Repairable fatigue damage
Brittle fracture	Corrosion

The following must be satisfied:

$$\text{Specified loads} \times \gamma f \text{ (load factor)} \leqslant \frac{\text{Material strength}}{\gamma m \text{ (material factor)}}$$

where $\gamma m = 1.0$

Values of the load factor are summarised in Table 1.12.

Table 1.12 BS 5950 Load factors γf and combinations

Loading	Load factor γf
Dead load	1.4
Dead load restraining uplift or overturning	1.0
Dead load acting with wind and imposed loads combined	1.2
Imposed loads	1.6
Imposed load acting with wind load	1.2
Wind load	1.4
Wind load acting with imposed load or crane load	1.2
Forces due to temperature effects	1.2
Crane loading effects	
Vertical load	1.6
Vertical load acting with horizontal loads (crabbing or surge)	1.4
Horizontal load	1.6
Horizontal load acting with vertical	1.4
Crane load acting with wind load*	1.2

* When considering wind or imposed load and crane loading acting together the value of γf for dead load may be taken as 1.2.

For the ultimate limit state of fatigue and all serviceability limit states $\gamma f = 1.0$

In this manual any load capacities given are in the terms of BS 5950 *ultimate* strength (i.e. Material strength γ m = 1.0), generally a function of the guaranteed yield stress of the material from BS 4360. They must be compared with factored working loads as given by Table 1.12 in satisfying compliance. If a working load is supplied then its appropriate proportions should be multiplied by the load factors from Table 1.12. As an approximation a working load can be multiplied by an averaged load factor of say 1.5 if the contribution of dead and imposed loads are approximately equal.

1.12.2 Bridges

Bridges are designed and constructed to BS 5400 [3] which covers steel, concrete, composite construction and bearings. It is adopted by the main UK highway and railway bridge authorities. It has been widely accepted in other countries and used as a model for other Codes. The UK Department of Transport implements BS 5400 with its own standards which in some cases vary with individual Code clauses. In particular the intensity of highway loading is increased to reflect the higher proportion of heavy commercial vehicles using UK highways since publication of the code.

BS 5400 uses a limit state concept similar to BS 5950. Many of the strength formulae are similar but there are additional clauses dealing with, for example, longitudinally stiffened girders, continuous composite beams and fatigue. In BS 5400 the breakdown of partial safety factors and the assessment of material strengths are different so that any capacities given in this book, where applicable to bridges, should not be used other than as a rough guide.

2 DETAILING PRACTICE

2.1 GENERAL

Drawings of steelwork whether engineer's drawings or workshop drawings should be carried out to a uniformity of standard to minimise the possible source of errors. Individual companies will have particular requirements suited to their own operation, but the guidance given herein is intended to reflect good practice. Certain conventions such as welding symbols are established by a Standard or other Code and should be used wherever possible.

2.2 LAYOUT OF DRAWINGS

Drawing sheet sizes should be standardised. BS 3429 [21] gives the international 'A' series, but many offices use the 'B' series. Typical sizes used are shown in Table 2.1.

Table 2.1 Drawing sheet sizes

Designation	Size mm	Main purpose
A0*	1189 × 841	Arrangement drawings
A1*	841 × 594	Detailed drawings
A2	594 × 420	Detailed drawings
A3*	420 × 297	Sketch sheets
A4*	297 × 210	Sketch sheets
B1	1000 × 707	Detailed drawings

* Widely used.

All drawings must contain a title block including company name, columns for the contract name/number, client, drawing number, drawing title, drawn/checked signatures, revision block, and notes column. Notes should, as far as possible, all be in the notes column. Figure 2.1 shows typical drawing sheet information.

2.3 LETTERING

No particular style of lettering is recommended but the objective is to provide, with reasonable rapidity, distinct uniform letters and figures that will ensure they can be read easily and produce legible copy prints. Faint guide lines should be used and trainee detailers and engineers should be taught to practice the art of printing which, if neatly executed, increases user confidence. Experienced detailers merely use a straight edge placed below the line when lettering.

The minimum size is 2.5 mm bearing in mind that microfilming or other reductions may be made. Stencils should not be necessary but may be used for view of drawing titles which should be underlined. (Note: stencilled lettering is shown in this manual merely for the sake of uniformity.) Underlining of other lettering should not be done except where special emphasis is required. Punctuation marks should not be used unless essential to the sense of the note.

2.4 DIMENSIONS

Arrow heads should have sharp points, touching the lines to which they refer. Dimension lines should be thin but full lines stopped just short of the detail. Dimension figures should be placed immediately above the dimension line and near its centre. The figures should be parallel to the line, arranged so that they can be *read from the bottom or right hand side* of the drawing. Dimensions should normally be given in millimetres and accurate to the nearest whole millimetre.

2.5 PROJECTION (See figure 2.2)

Third angle projection should be used whenever possible. With this convention each view is so placed that it represents the side of the object nearest to it in the adjacent view. The notable exception is the base detail on a column, which by convention is shown as in figure 6.1.5.

2.6 SCALES

Generally scales as follows should be used:
   1:5, 1:10, 1:20, 1:25, 1:50, 1:100, 1:200.
Scales should be noted in the title block, and not normally repeated in views. Beams, girders, columns and bracings should preferably be drawn true scale, but may exceptionally be drawn to a smaller longitudinal scale. The section depth and details and other connections must be drawn to scale and in their correct relative positions. A series of sections through a member should be to the same scale, and preferably be arranged in line, in correct sequence.

For bracing systems, lattice girders and trusses a convenient practice is to draw the layout of the centre lines of members to one scale and superimpose details to a larger scale at intersection points and connections.

2.7 REVISIONS

All revisions must be noted on the drawing in the revision column and every new issue is identified by a date and issue letter (see figure 2.1).

2.8 BEAM AND COLUMN DETAILING CONVENTION
(See figure 2.2)

When detailing columns from a floor plan two main views, A viewed from the bottom and B from the right of the plan, must always be given. If necessary, auxiliary views must be added to give the details on the other sides.

Whenever possible columns should be detailed vertically on the drawing, but often it will be more efficient to draw horizontally in which case the base end must be at the right hand side of the drawing with view A at the bottom and view B at the top. If columns are detailed vertically the base will naturally be at the bottom with view A on the left of the drawing and view B at the right. Auxilliary views are drawn as necessary. An example of a typical column detail is shown in figure 6.1.5.

When detailing a beam from a floor plan, the beam must always be viewed from the bottom or right of the plan. If a beam connects to a seating, end connections must be dimensioned from the bottom flange upwards but if connected by other means (e.g. web cleats, end plates) then end connections must be dimensioned from top flange downwards (see figure 6.1.4).

Holes in flanges must be dimensioned from centre-line of web. Rolled steel angles (RSA), channels, etc. should when possible be detailed with the outstanding leg on farside with 'backmark' dimension given to holes.

2.9 ERECTION MARKS

An efficient and simple method of marking should be adopted and each loose member or component must have a separate mark. For beam/column structures the allocation of marks for members is shown in figure 2.2.

On beams the mark should be located on the top flange at the north or east (right-hand) end. On columns the mark should be located on the lower end of the shaft on the flange facing north or east. On vertical bracings the mark should be located at the lower end.

To indicate on a detail drawing where an erection mark is to be painted, the word *mark* contained in a rectangle shall be shown on each detail with an arrow pointing to the position required.

Care should be taken when marking weathering steel to ensure it does not damage finish or final appearance.

2.10 OPPOSITE HANDING

Difficulties frequently arise in both drawing offices and workshops over what is meant by the term *opposite hand*.

Members which are called off on drawings as '1 As Drawn, 1 Opp. Hand' are simply pairs or one right hand and one left hand. A simple illustration of this is the human hand. The left hand is opposite hand to the right hand and vice-versa. Any steelwork item must always be opposite handed about a longitudinal centre or datum line and never from end to end. Figure 2.2 shows an example of calling off to opposite hand, with the item referred to also shown to illustrate the principle.

Erection marks are usually placed at the east or north end of an item and opposite handing does not alter this. The erection mark must stay in the position shown on the drawing, i.e. the erection mark is not handed.

2.11 WELDS

Welds should be identified using weld symbols as shown in figure 4.4 and should not normally be drawn in elevation using 'whiskers' or in cross section. In particular cases it may be necessary to draw weld cross sections to enlarged scale showing butt weld edge preparations such as for complex joints including cruciform type. Usual practice is for workshop butt weld preparations to be shown on separate *weld procedure sheets* not forming part of the drawings. Site welds should be detailed on drawings with the dimensions taking into account allowances for weld shrinkage at site. Space should be allowed around the weld whenever possible so as to allow downhand welding to be used.

2.12 BOLTS

Bolts should be indicated using symbolic representation as in figure 2.2 and should only be drawn with actual bolt and nut where necessary to check particularly tight clearances.

2.13 HOLDING DOWN BOLTS

A typical holding down bolt detail should be drawn out defining length, protrusion above baseplate, thread length, anchorage detail pocket and grouting information and other HD bolts described by notes or schedules.

Typical notes are as follows which could be printed onto a drawing or issued separately as a specification.

Notes on holding down bolts

(1) HD bolts shall be cast into foundations using template, accurately to line and level within pockets of size shown to permit tolerance. Immediately after concreting in all bolts shall be 'waggled' to ensure free movement.

(2) Temporary packings used to support and adjust steelwork shall be suitable steel shims placed concentrically with respect to the baseplate. If to be left in place, they shall be positioned such that they are totally enclosed by 30 mm minimum grout cover.

(3) No grouting shall be carried out until a sufficient portion of the structure has been finally adjusted and secured. The spaces to be grouted shall be clear of all debris and free water.

(4) Grout shall have a characteristic strength not less than that of the surrounding concrete nor less than 20 N/mm^2. It shall be placed by approved means such that the spaces around HD bolts and beneath the baseplate are completely filled.

(5) Baseplates greater than 400 mm wide shall be provided with at least 2 grout holes preferably not less than 30 mm diameter.

(6) Washer plates or other anchorages for securing HD bolts shall be of sufficient size and strength. They shall be designed so that they prevent pull-out failure. The concrete into which HD bolts are anchored shall be reinforced with sufficient overlap and anchorage length so that uplift forces are properly transmitted.

2.14 ABBREVIATIONS

It is economic to use abbreviations in using space economically on drawings. A list of suitable abbreviations is given in Table 2.2.

Table 2.2 List of abbreviations

Description	Abbreviate on drawings	Description	Abbreviate on drawings
Overall length	O'ALL	Girder	GDR
Unless otherwise stated	UOS	Column	COL
Diameter	DIA or Φ	Beam	BEAM
Long	LG	High strength friction grip bolts	HSFG BOLTS
Radius	r or RAD	24 mm diameter bolts grade 8.8	M24 (8.8) BOLTS
Vertical	VERT	Countersunk	CSK
Mark	MK	Full penetration butt weld	FPBW
Dimension	DIM	British Standard BS 4360: 1986	BS 4360: 86
Near side, far side	N SIDE F SIDE	100 mm length × 19 diameter shear studs	100 × 19 SHEAR STUDS
Opposite hand	OPP HAND	Plate	PLT
Centre to centre	C/C	Bearing plate	BRG PLT
Centre-line	℄	Packing plate	PACK
Horizontal	HORIZ	Gusset plate	GUSSET
Drawing	DRG	30 mm diameter holding down bolts grade 8.8, 600 mm long	M30 (8.8) HD BOLTS 600 LG
Not to scale	NTS		
Typical	TYP	Flange plate	FLG
Nominal	NOM	Web plate	WEB
Reinforced concrete	RC	Intermediate stiffener	STIFF
Floor level	FL	Bearing stiffener	BRG STIFF
Setting out point	SOP	Fillet weld	FW (but use welding symbols!)
Required	REQD	Machined surface	m/c ↓
		Fitted to bear	FIT
Section A–A	A–A	Cleat	CLEAT
Right Angle	90°	35 pitches at 300 centres = 10500	35 × 300 c/c = 10500
45 degrees	45°		
Slope 1:20		70 mm wide × 12 mm thick plate	70 × 12 PLT
20 number required	20 No	120 mm wide × 10 mm thick × 300 mm long plate	120 × 10 PLT × 300
203 × 203 × 52 kg/m universal column	203 × 203 × 52 UC	25 mm thick	25 THK
406 × 152 × 60 kg/m universal beam	406 × 152 × 60 UB	80 mm × 80 mm plate × 6 mm thick	80 SQ × 6 PLT
150 × 150 × 10 mm angle	150 × 150 × 10 RSA (or L)		
305 × 102 channel	305 × 102 ⊏ or 305 × 102 CHANN		
127 × 114 × 29.76 kg/m joist	127 × 114 × 29.76 JOIST		
152 × 152 × 36 kg/m structural tee	152 × 152 × 36 TEE		

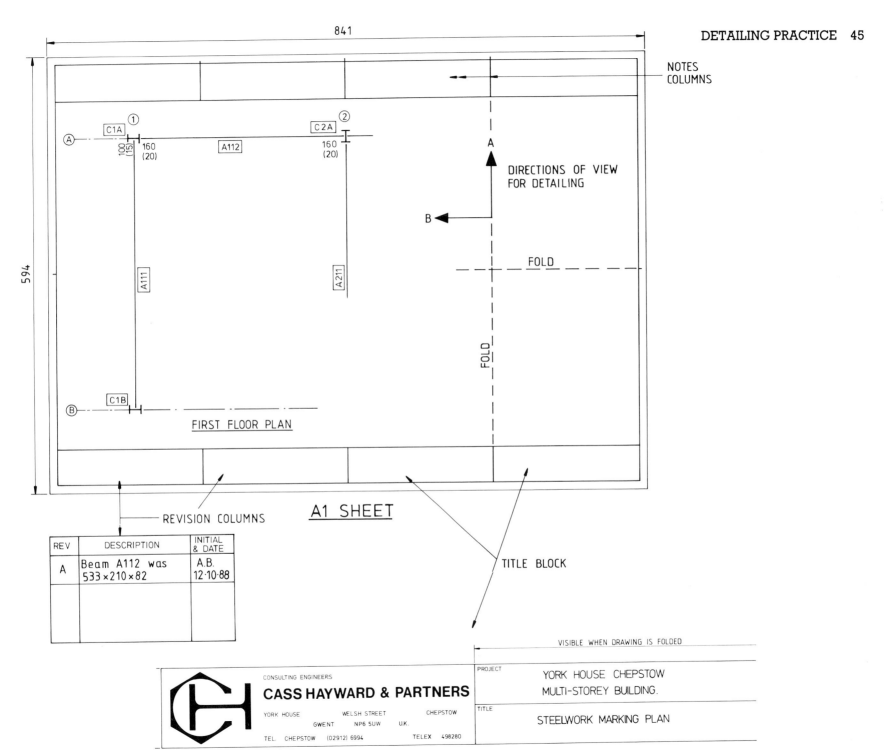

Figure 2.1 Drawing sheets and marking system

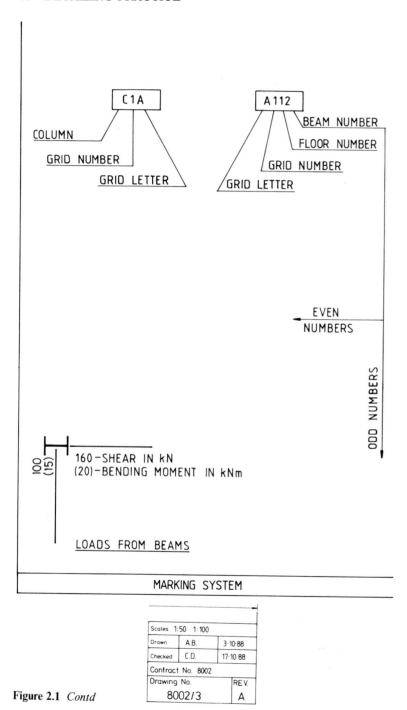

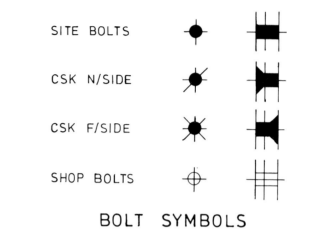

BOLT SYMBOLS

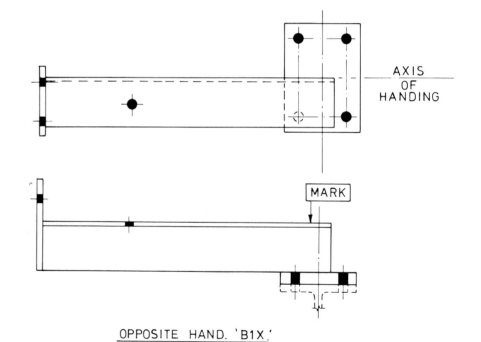

OPPOSITE HAND. 'B1X.'
(SHOWS HANDED BRACKET)

OPPOSITE HANDING

Figure 2.1 *Contd*

Scales	1:50 1:100	
Drawn	A.B.	3·10·88
Checked	C.D.	17·10·88
Contract No. 8002		
Drawing No.		REV.
8002/3		A

Figure 2.2 Dimensioning and conventions

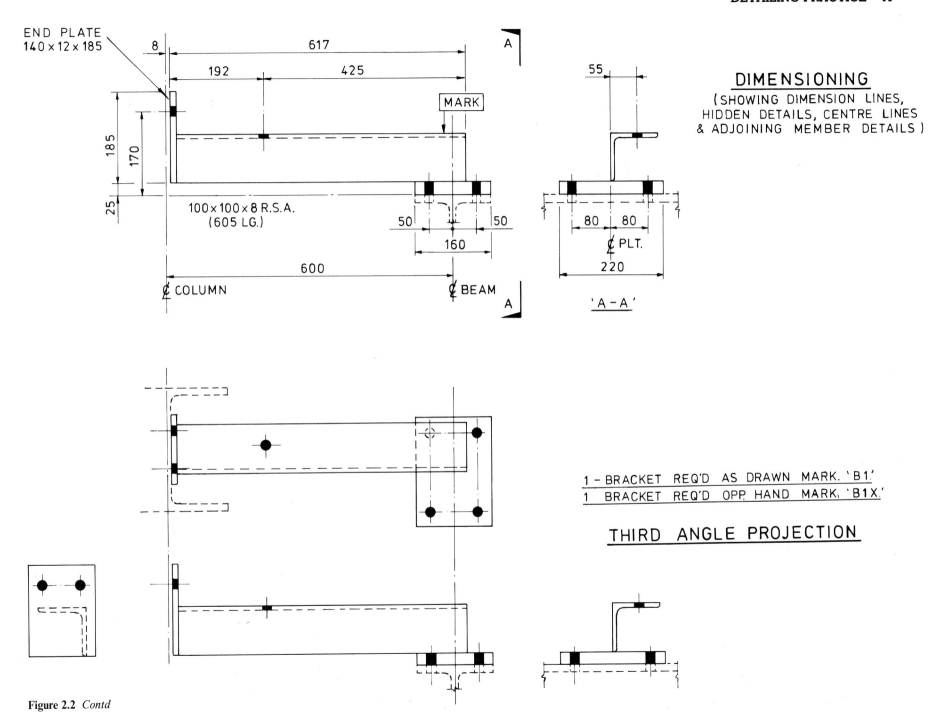

END PLATE
140 × 12 × 185

100 × 100 × 8 R.S.A.
(605 LG.)

MARK

₵ COLUMN

₵ BEAM

'A - A'

DIMENSIONING

(SHOWING DIMENSION LINES,
HIDDEN DETAILS, CENTRE LINES
& ADJOINING MEMBER DETAILS)

1 - BRACKET REQ'D AS DRAWN MARK. 'B1.'
1 BRACKET REQ'D OPP. HAND MARK, 'B1X.'

THIRD ANGLE PROJECTION

Figure 2.2 *Contd*

3 DESIGN GUIDANCE

3.1 GENERAL

Limited design guidance is included in this manual for selecting *simple connections* and *simple baseplates* which can be carried out by the detailer without demanding particular design skills. Other connections including *rigid connections* and the design of members such as beams, girders, columns, bracings and lattice structures will require specific design calculations. Load capacities for members are contained in the Design Guide to BS 5950 [18] and from other literature as given in the Further Reading.

Capacities of bolts and welds to BS 5950 are included in Tables 3.1, 3.2 and 3.3 so that detailers can proportion elementary connections such as welds and bolts to gusset plates etc.

3.2 LOAD CAPACITIES OF SIMPLE CONNECTIONS

Ultimate load capacities for a range of simple web angle cleat/end plate type beam/column and beam/beam connections for universal beams are given in Tables 3.1, 3.2 and 3.3. The capacities must be compared with *factored* loads to BS 5950. The tables indicate whether bolt shear, bolt bearing, web shear or weld strength are critical so that different options can be examined. The range of coverage is listed at the foot of this page.

Capacities in kN are presented under the following symbols:

Connection to beam   Bc – RSA cleats
                             Be – End plates

Connection to column   S1 – one sided connection – maximum
                             S2 – two sided connection – total reaction from 2 incoming beams sharing the same bolt group

Table	Steel grade	M20 bolts grade	Grade 43 RSA web cleats		Grade 43 end plates		Number of bolts rows to column/ beam	Welds to end plate	
			To column	To beam	To columns	To beams		N11 to N6	N5 to N1
3.1	43	4.6	$100 \times 100 \times 10$	$90 \times 90 \times 10$	200×10	160×8	Range	8 mm	6 mm
3.2	43	8.8	$100 \times 100 \times 10$	$90 \times 90 \times 10$	200×10	160×10	N11 to N1	fillet	fillet
3.3	50	8.8	$100 \times 100 \times 10$	$90 \times 90 \times 10$	200×10	160×10		welds	welds

Table 3.1 Simple connections, bolts grade 4.6, members grade 43. See figure 3.1

Thickness (mm) of beam web or column web/flange connected

Type	UB sizes for beam	Symbols	4		6		8		10		12		14		16		18		20	
N11	914 × 419 914 × 305	Bc Be	–	–	–	–	–	–	–	–	–	–	803	1714	803	*1810*	803	*1810*	803	1803
		S1 S2	–	–	–	–	–	–	862	*1531*	862	1725	862	1725	862	1725	862	1725	862	1725
N10	914 × 419 914 × 305 838 × 292	Bc Be	–	–	–	–	–	–	–	–	–	–	721	1558	721	*1642*	721	*1642*	721	*1642*
		S1 S2	–	–	–	–	784	*1392*	784	1568	784	1568	784	1568	784	1568	784	1568	784	1568
N9	914 × 419 914 × 305 838 × 292 762 × 267	Bc Be	–	–	–	–	–	–	–	–	638	1202	638	*1474*	638	*1474*	638	*1474*	638	1474
		S1 S2	–	–	–	–	706	*1253*	706	1411	706	1411	706	1411	706	1411	706	1411	706	1411
N8	914 × 419 914 × 305 838 × 292 762 × 267 686 × 254	Bc Be	–	–	–	–	–	–	556	890	556	1068	556	1247	556	*1306*	556	*1306*	556	1306
		S1 S2	–	–	–	–	627	*835*	627	*1114*	627	1254	627	1254	627	1254	627	1254	627	1254
N7	838 × 292 762 × 267 686 × 254 610 × 305 610 × 229	Bc Be	–	–	–	–	–	–	473	779	473	935	473	1091	473	*1138*	473	*1138*	473	1138
		S1 S2	–	–	–	–	549	*731*	549	*974*	549	1098	549	1098	549	1098	549	1098	549	1098
N6	686 × 254 610 × 305 610 × 229 533 × 210	Bc Be	–	–	–	–	*339*	554	390	668	390	801	390	935	390	*970*	390	*970*	390	970
		S1 S2	–	–	470	*626*	470	*835*	470	941	470	941	470	941	470	941	470	941	470	941
N5	457 × 191 457 × 152 406 × 178 406 × 140	Bc Be	–	–	*200*	347	267	462	314	557	314	*610*	–	–	–	–	–	–	–	–
		S1 S2	*348*	*348*	392	*522*	392	*696*	392	784	392	784	392	784	392	784	392	784
N4	457 × 191 457 × 152 406 × 178 406 × 140 356 × 171 356 × 127	Bc Be	98	185	*147*	277	*196*	370	228	445	228	*484*	–	–	–	–	–	–	–	–
		S1 S2	*278*	*278*	314	*418*	314	*557*	314	627	314	627	314	627	314	627	314	627
N3	406 × 178 406 × 140 356 × 171 356 × 127 305 × 165 305 × 127 254 × 146 254 × 102	Bc Be	65	139	*97*	208	*129*	277	152	347	–	–	–	–	–	–	–	–	–	–
		S1 S2	*209*	*209*	235	*313*	235	*418*	235	470	235	*470*	235	470	235	470	235	470
N2	305 × 165 305 × 127 254 × 146 305 × 102 203 × 133 254 × 102 203 × 102	Bc Be	35	92	*53*	139	*70*	185	84	231	–	–	–	–	–	–	–	–	–	–
		S1 S2	*139*	*139*	157	*209*	157	*278*	157	314	157	*314*	157	314	157	314	157	314
N1	254 × 146 254 × 102 203 × 133	Bc Be	35	92	*53*	139	*70*	185	84	231	–	–	–	–	–	–	–	–	–	–
		S1 S2	*70*	*70*	78	*104*	78	*139*	78	157	78	157	78	157	78	157	78	157

Connection to beam
Bc – RSA cleats – capacity in kN – lesser of bolt shear and bolt bearing to web. Value in italics is bolt bearing where less.
Be – End plates – capacity in kN – lesser of web shear or weld strength. Value in italics is weld strength where less.

Connection to column
S1 – RSA cleats or end plates – capacity in kN
S2 – RSA cleats or end plates – capacity in kN
Least of bolt shear, bolt bearing to web, or bolt bearing to cleat. Value in italics is bolt bearing where the least.

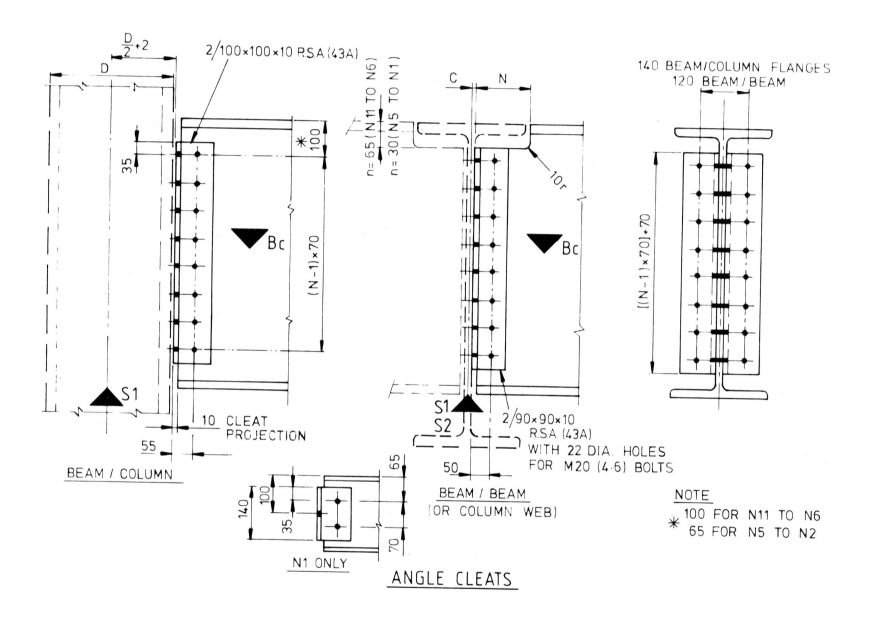

Figure 3.1 Simple connections

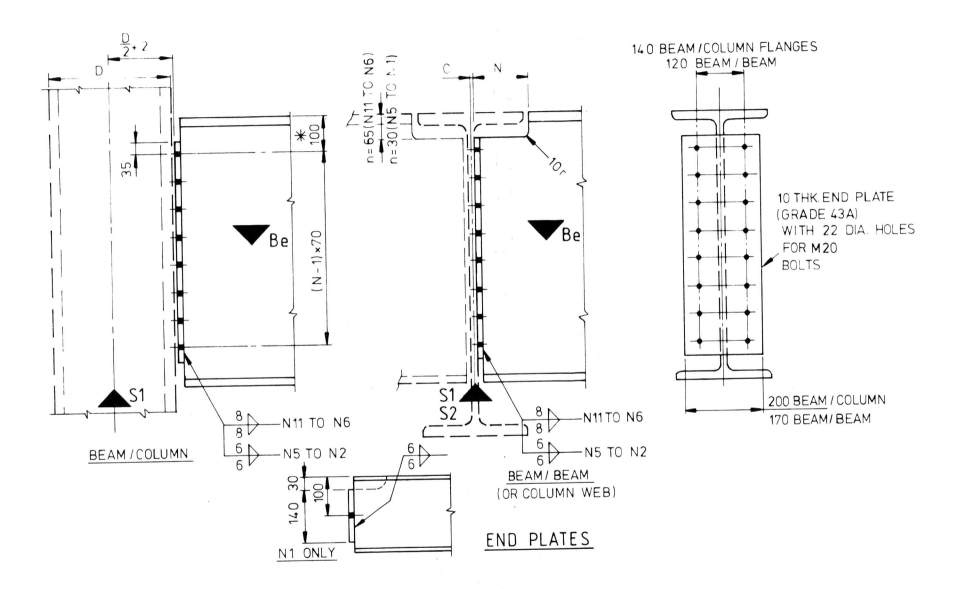

END PLATES

Figure 3.1 *Contd*

Table 3.2 Simple connections, bolts grade 8.8, members grade 43. See figure 3.1

Type	UB sizes for beam	Symbols	4	4	6	6	8	8	10	10	12	12	14	14	16	16	18	18	20	20
			\<--- Thickness (mm) of beam web or column web/flange connected ---\>																	
N11	914 × 419, 914 × 305	Bc Be	–	–	–	–	–	–	–	–	–	–	1304	1714	1490	1810	1676	1810	1862	1810
		S1 S2	–	–	–	–	1619	1619	**2001**	2024	**2001**	2429	**2001**	2834	**2001**	3238	**2001**	3643	**2001**	4002
N10	914 × 419, 914 × 305, 838 × 292	Bc Be	–	–	–	–	–	–	–	–	–	–	1169	1558	1336	1642	1503	1642	1670	1642
		S1 S2	–	–	–	–	1472	1472	**1817**	1840	**1817**	2208	**1817**	2576	**1817**	2944	**1817**	3312	**1817**	3634
N9	914 × 419, 914 × 305, 838 × 292, 762 × 267	Bc Be	–	–	–	–	–	–	–	–	886	1202	1034	1474	1182	1474	1329	1474	1477	1474
		S1 S2	–	–	–	–	1325	1325	**1633**	1656	**1633**	1987	**1633**	2318	**1633**	2650	**1633**	2980	**1633**	3266
N8	914 × 419, 914 × 305, 838 × 292, 762 × 267, 686 × 254	Bc Be	–	–	–	–	–	–	642	890	770	1068	898	1247	1027	1306	1155	1306	1283	1306
		S1 S2	–	–	883	883	1178	1178	**1449**	1472	**1449**	1766	**1449**	2061	**1449**	2355	**1449**	2650	**1449**	2898
N7	838 × 292, 762 × 267, 686 × 254, 610 × 305, 610 × 229	Bc Be	–	–	–	–	–	–	545	779	654	935	763	1091	872	1138	981	1138	1090	1138
		S1 S2	–	–	773	773	1030	1030	**1265**	1288	**1265**	1546	**1265**	1803	**1265**	2061	**1265**	2318	**1265**	2530
N6	686 × 254, 610 × 305, 610 × 229, 533 × 210	Bc Be	–	–	–	–	359	554	448	668	457	801	628	935	717	970	807	970	897	970
		S1 S2	–	–	663	663	883	883	**1081**	1104	**1081**	1325	**1081**	1546	**1081**	1766	**1081**	1987	**1081**	2162
N5	457 × 191, 457 × 152, 406 × 178, 406 × 140	Bc Be	–	–	212	347	282	462	353	557	423	610	–	–	–	–	–	–	–	–
		S1 S2	368	368	552	552	736	736	**897**	920	697	1104	**897**	1288	**897**	1472	**897**	1656	**897**	1794
N4	457 × 191, 457 × 152, 406 × 178, 406 × 140, 356 × 171, 356 × 127	Bc Be	104	185	156	277	208	370	260	445	311	484	–	–	–	–	–	–	–	–
		S1 S2	294	294	442	442	589	589	**713**	736	**713**	883	**713**	1030	**713**	1178	**713**	1325	**713**	1426
N3	406 × 178, 406 × 140, 356 × 171, 356 × 127, 305 × 165, 305 × 127, 254 × 146, 254 × 102	Bc Be	68	139	103	208	137	277	171	347	–	–	–	–	–	–	–	–	–	–
		S1 S2	221	221	331	331	442	442	**529**	552	**529**	662	**529**	773	**529**	883	**529**	994	**529**	1058
N2	305 × 165, 305 × 127, 254 × 146, 305 × 102, 203 × 133, 254 × 102, 203 × 102	Bc Be	37	92	56	139	74	185	93	231	–	–	–	–	–	–	–	–	–	–
		S1 S2	147	147	221	221	294	294	**345**	368	**345**	442	**345**	515	**345**	589	**345**	662	**345**	690
N1	254 × 146, 254 × 102, 203 × 133	Bc Be	37	92	56	139	74	185	93	231	–	–	–	–	–	–	–	–	–	–
		S1 S2	74	74	110	110	147	147	**161**	184	**161**	221	**161**	258	**161**	294	**161**	331	**161**	322

Connection to beam — Bc – RSA cleats – capacity in kN – lesser of bolt shear and bolt bearing to web. Value in italics is bolt bearing where less.

Be – End plates – capacity in kN – lesser of web shear or weld strength. Value in italics is weld strength where less.

Connection to column — S1 – RSA cleats or end plates – capacity in kN. Least of bolt shear, bolt bearing to web, or bolt bearing to cleat. Value in italics is bolt bearing where the least. Value shown in bold type if bolt bearing to cleats is least.

S2 – RSA cleats or end plates – capacity in kN.

Table 3.3 Simple connections, bolts grade 8.8, members grade 50. See figure 3.1

Thickness (mm) of beam web or column web/flange connected

Type	UB sizes for beam	Symbol		4		6		8		10		12		14		16		18		20	
N11	914 × 419 914 × 305	Bc	Be	–	–	–	–	–	–	–	–	–	–	*1558*	*1810*	*1781*	*1810*	*1862*	*1810*	*1862*	*1810*
		S1	S2	–	–	–	–	*1935*	*1935*	**2001**	*2420*	**2001**	*2904*	**2001**	*3388*	**2001**	*3872*	**2001**	*4002*	**2001**	*4002*
N10	914 × 419 914 × 305 838 × 292	Bc	Be	–	–	–	–	–	–	–	–	–	–	*1398*	*1642*	*1597*	*1642*	*1670*	*1642*	*1670*	*1642*
		S1	S2	–	–	–	–	*1760*	*1760*	**1817**	*2200*	**1817**	*2640*	**1817**	*3080*	**1817**	*3520*	**1817**	*3634*	**1817**	*3634*
N9	914 × 419 914 × 305 838 × 292 762 × 267	Bc	Be	–	–	–	–	–	–	–	–	*1060*	*1474*	*1236*	*1474*	*1413*	*1474*	*1477*	*1474*	*1477*	*1474*
		S1	S2	–	–	–	–	*1584*	*1584*	**1633**	*1980*	**1633**	*2376*	**1633**	*2772*	**1633**	*3168*	**1633**	*3266*	**1633**	*3266*
N8	914 × 419 914 × 305 838 × 292 762 × 267 686 × 254	Bc	Be	–	–	–	–	–	–	*768*	*1142*	*921*	*1306*	*1074*	*1306*	*1228*	*1306*	*1283*	*1306*	*1283*	*1306*
		S1	S2	–	–	*1056*	*1056*	*1408*	*1408*	**1449**	*1760*	**1449**	*2112*	**1449**	*2464*	**1449**	*2816*	**1449**	*2898*	**1449**	*2898*
N7	838 × 292 762 × 267 686 × 254 610 × 305 610 × 229	Bc	Be	–	–	–	–	–	–	*652*	*1000*	*782*	*1138*	*912*	*1138*	*1043*	*1138*	*1090*	*1138*	*1090*	*1138*
		S1	S2	–	–	*924*	*924*	*1231*	*1231*	**1265**	*1540*	**1265**	*1848*	**1265**	*2156*	**1265**	*2464*	**1265**	*2530*	**1265**	*2530*
N6	686 × 254 610 × 305 610 × 229 533 × 210	Bc	Be	–	–	–	–	*429*	*716*	*536*	*857*	*643*	*970*	*751*	*970*	*858*	*970*	*897*	*970*	*897*	*970*
		S1	S2	–	–	*792*	*792*	*1056*	*1056*	**1081**	*1320*	**1081**	*1584*	**1081**	*1848*	**1081**	*2112*	**1081**	*2162*	**1081**	*2162*
N5	457 × 191 457 × 152 406 × 178 406 × 140	Bc	Be	–	–	253	447	337	596	422	610	506	610	–	–	–	–	–	–	–	–
		S1	S2	*440*	*440*	*660*	*660*	*880*	*880*	**897**	*1100*	**897**	*1320*	**897**	*1540*	**897**	*1760*	**897**	*1794*	**897**	*1794*
N4	457 × 191 457 × 152 406 × 178 406 × 140 356 × 171 356 × 127	Bc	Be	124	239	186	358	248	477	*310*	484	372	484	–	–	–	–	–	–	–	–
		S1	S2	*352*	*352*	*528*	*528*	*704*	*704*	**713**	*880*	**713**	*1056*	**713**	*1232*	**713**	*1408*	**713**	*1426*	**713**	*1426*
N3	406 × 178 406 × 140 356 × 171 356 × 127 305 × 165 305 × 127 254 × 146 254 × 102	Bc	Be	82	179	123	268	164	358	205	358	–	–	–	–	–	–	–	–	–	–
		S1	S2	*264*	*264*	*396*	*396*	*528*	*528*	**529**	*660*	**529**	*792*	**529**	*924*	**529**	*1056*	**529**	*1058*	**529**	*1058*
N2	305 × 165 305 × 127 254 × 146 305 × 102 203 × 133 254 × 102 203 × 102	Bc	Be	44	119	66	179	89	231	*111*	231	–	–	–	–	–	–	–	–	–	–
		S1	S2	*176*	*176*	*264*	*264*	**345**	*352*	**345**	*440*	**345**	*528*	**345**	*666*	**345**	*690*	**345**	*690*	**345**	*690*
N1	254 × 146 254 × 102 203 × 133	Bc	Be	44	119	66	179	89	231	111	231	–	–	–	–	–	–	–	–	–	–
		S1	S2	*88*	*88*	*132*	*132*	**161**	*176*	**161**	*220*	**161**	*264*	**161**	*308*	**161**	*322*	**161**	*322*	**161**	*322*

Connection to beam
Bc – RSA cleats – capacity in kN – lesser of bolt shear and bolt bearing to web. Value in italics is bolt bearing where less.
Be – End plates – capacity in kN – lesser of web shear or weld strength. Value in italics is weld strength where less.

Connection to column
S1 – RSA cleats or end plates – capacity in kN
S2 – RSA cleats or end plates – capacity in kN
Least of bolt shear, bolt bearing to web, or bolt bearing to cleat. Value in italics is bolt bearing where the least. Value shown in bold type if bolt bearing to cleats is least.

Worked example
The following example illustrates use of Tables 3.1, 3.2 and 3.3:

Question
A beam of size 686 × 254 × 140 UB in grade 43 steel has a factored end reaction of 750 kN. Design the connection using RSA web cleats:

(a)   To a perimeter column size 305 × 305 × 97 uc, of grade 43 steel via its flange.

(b)   To a similar internal column via its web forming a two sided connection with another beam having the same reaction.

Answer
(a)   To perimeter columns
Connection to beam:
686 × 254 × 140 UB – web thickness 12.4 mm
From Table 3.1 (grade 4.6 bolts) maximum value of Bc = 556 kN for N8 type which is insufficient. Capacity cannot be increased by thicker webbed beam because bolt shear governs (because value is not in italics).
So try grade 8.8 bolts:
From Table 3.2 value of Bc = 770 kN for 12 mm web for N8 type
                               = 898 kN for 14 mm web for N8 type
Interpolation for 12.4 mm web gives Bc = 796 kN > 750 ACCEPT
Connection to column: 305 × 305 × 97 UC – flange thickness 15.4 mm
From Table 3.2 value of S1 = 1449 kN for 14 mm flange
                               = 1449 kN for 16 mm flange
               Therefore S1 = 1449 kN for 15.4 mm flange > 750
                               ACCEPT
Therefore connection is N8 with 100 × 100 × 10 RSA cleats, i.e. 8 rows of M20 (8.8) bolts.
(b)   To internal column connection to beam
Connection to beam
As for (a) i.e. N8 type using grade 8.8 bolts.
Connection to column: 305 × 305 × 97 UC – web thickness 9.9 mm.
From Table 3.2, value of S2 = 1178 kN for 8 mm web
                               = 1472 kN for 10 mm web
Interpolation for 9.9 mm web gives S2 = 1457 kN < 2 × 750 = 1500 kN
Therefore insufficient, but note that bolt bearing is critical (because value is in italics) so try grade 50 steel for column.

From Table 3.3, value of S2 = 1408 kN for 8 mm web
                               = 1760 kN for 10 mm web
Interpolation for 9.9 mm web gives S2 = 1742 kN > 1500 ACCEPT
Alternatively try larger diameter bolts:

For M22 (8.8) bolt:
From Table 3.5: giving capacities of single bolts:
double shear value = 227 kN
bearing to 2/10 mm (43) cleats 2 × 101 = 202 kN
bearing to UB web (43)
       9 mm thick   91 kN
       10 mm thick   101 kN
Interpolation for 9.9 mm thick gives 100 kN
therefore bearing to UC web governs
So capacity is 16 bolts × 100 = 1600 kN > 1500 ACCEPT
Therefore M22 (8.8) bolts can be used instead of using grade 50 steel for the column.

3.3 SIZES AND LOAD CAPACITY OF SIMPLE COLUMN BASES

Ultimate capacities and baseplate thicknesses using grade 43 steel for a range of simple square column bases with universal column or square hollow section columns are given in Table 3.4. These capacities must be compared with factored loads to BS 5950.
Baseplate thickness is derived to BS 5950: Part 1 clause 4.13.2.2:

$$t = \left[\frac{2.5}{pyp} w \, (a^2 - 0.3b^2) \right]^{1/2}$$

where a = greater projection of plate in mm
      b = lesser projection of plate in mm
      w = pressure on underside of plate in N/mm²
    pyp = yield strength of plate in N/mm²
         (which shall not be taken as greater than 270 N/mm²).

Worked example
Question
The following example illustrates use of Table 3.4:
A 305 × 305 × 97 UC column carries a factored vertical load of 3000 kN at the base. The foundation concrete has an ultimate strength of 30N/mm². Select a baseplate size.

Answer
From Table 3.4 width of base for concrete strength 30 N/mm² is 500 mm for P = 3000 kN.
Thickness = 30 mm
Therefore baseplate minimum size is 500 × 30 × 500 in grade 43 steel.

Table 3.4 Simple column bases (see figure 3.2 opposite)

WIDTH OF BASE B (mm)

UC	SHS	200	225	250	275	300	325	350	375	400	425	450	475	500	525	550	575	600	625	650	675	700	725	750	Concrete Strength f_{cu}-N/mm²
P (kN)		480	600	750	900	1080	1260	1470	1680	1920	2160	2430	2700	3000	3300	3630	3960	4320	4680	5070	5460	5880	6300	6750	
–	100	15	20	25	25	30	35	–	–	–	–	–	–	–	–	–	–	–	–	–	–	–	–	–	
152 × 152	150	–	–	15	20	25	25	30	35	40	40	45	50	55	–	–	–	–	–	–	–	–	–	–	
203 × 203	200	–	–	–	–	15	20	25	25	30	35	40	40	45	50	55	55	60	–	–	–	–	–	–	30
254 × 254	250	–	–	–	–	–	–	15	15	25	25	30	30	40	40	45	50	55	55	60	65	–	–	70	
305 × 305	300	–	–	–	–	–	–	–	–	15	20	25	25	30	35	40	40	45	50	55	55	60	65	70	
356 × 368	350	–	–	–	–	–	–	–	–	–	–	15	20	25	25	30	35	40	40	45	45	55	55	60	
356 × 406	400	–	–	–	–	–	–	–	–	–	–	–	–	15	20	25	25	30	35	40	40	45	50	55	

UC	RHS	200	225	250	275	300	325	350	375	400	425	450	475	500	525	550	575	600	625	650	675	700	725	750	Concrete Strength f_{cu}-N/mm²
P (kN)		640	810	1000	1210	1440	1690	1960	2250	2560	2890	3240	3610	4000	4410	4840	5290	5760	6250	6750	7290	7840	8410	9000	
–	100	20	25	25	30	35	–	–	–	–	–	–	–	–	–	–	–	–	–	–	–	–	–	–	
152 × 152	150	–	–	20	25	25	30	35	40	45	50	55	–	–	–	–	–	–	–	–	–	–	–	–	
203 × 203	200	–	–	–	–	20	25	25	30	35	40	45	50	55	55	60	–	–	–	–	–	–	–	–	40
254 × 254	250	–	–	–	–	–	–	20	25	25	30	35	40	45	50	55	55	60	65	75	75	–	–	–	
305 × 305	300	–	–	–	–	–	–	–	–	20	25	25	30	35	40	45	50	55	55	65	65	70	75	85	
356 × 368	350	–	–	–	–	–	–	–	–	–	–	20	25	25	30	35	40	45	50	55	55	60	65	80	
356 × 406	400	–	–	–	–	–	–	–	–	–	–	–	–	20	25	25	30	35	40	45	50	55	55	70	

UC	RHS	200	225	250	275	300	325	350	375	400	425	450	475	500	525	550	575	600	625	650	675	700	725	750	Concrete Strength f_{cu}-N/mm²
P (kN)		800	1010	1250	1510	1800	2110	2450	2810	3200	3610	4050	4510	5000	5510	6050	6610	7200	7810	8450	9110	9800	10510	11250	
–	100	20	25	30	35	–	–	–	–	–	–	–	–	–	–	–	–	–	–	–	–	–	–	–	
152 × 153	150	–	–	20	25	30	35	40	45	50	–	–	–	–	–	–	–	–	–	–	–	–	–	–	
203 × 203	200	–	–	–	–	20	25	30	35	40	45	50	55	–	–	–	–	–	–	–	–	–	–	–	50
254 × 254	250	–	–	–	–	–	–	20	25	30	35	40	45	50	55	60	65	70	75	–	–	–	–	–	
305 × 305	300	–	–	–	–	–	–	–	–	20	25	30	35	40	45	50	55	60	65	70	75	80	85	85	
356 × 368	350	–	–	–	–	–	–	–	–	–	–	20	25	30	35	40	45	50	55	60	65	70	75	80	
356 × 406	400	–	–	–	–	–	–	–	–	–	–	–	–	20	25	30	35	40	45	50	55	60	65	70	

The details are suitable for column bases comprising mainly vertical load only. The holding down bolts shown are suitable for nominal transverse or uplift forces and for securing/adjusting the steelwork during erection. A typical length is 24 diameter. Fixed bases (i.e. carrying bending moments) require larger HD bolts and increase in size of welds. Length of HD bolts must be adequate for anchorage and sufficient lapping reinforcement in the base supplied. Gusseted bases may be required in carrying loads in excess of those tabled or where bending moments are very significant. They are costly and simple column bases as shown should be used whenever possible.

P(kN). Ultimate vertical load capacity of concrete beneath base (stress 0.4 fcn) to B.S. 5950:Part 1:1985.

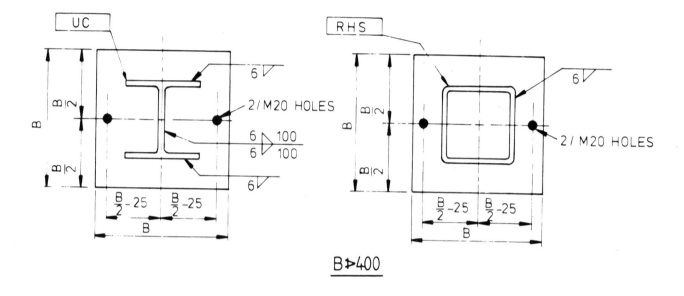

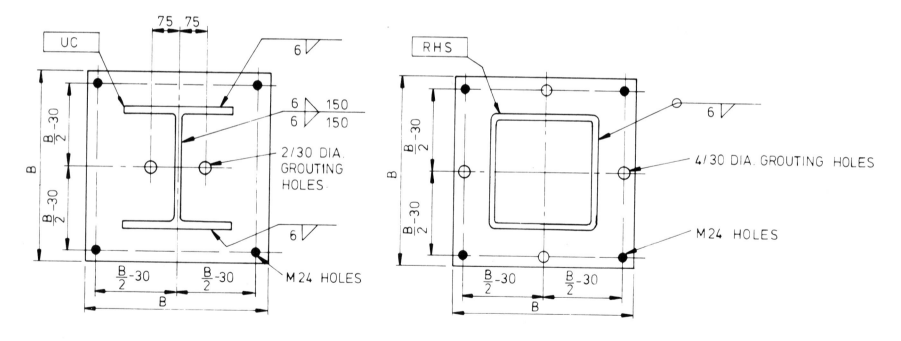

Figure 3.2 Simple column bases

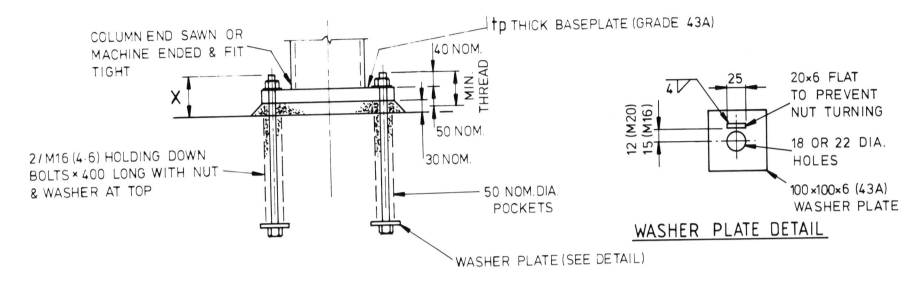

COLUMN END SAWN OR MACHINE ENDED & FIT TIGHT

tp THICK BASEPLATE (GRADE 43A)

X

2/M16 (4·6) HOLDING DOWN BOLTS × 400 LONG WITH NUT & WASHER AT TOP

40 NOM.

MIN. THREAD

50 NOM.

30 NOM.

50 NOM. DIA. POCKETS

WASHER PLATE (SEE DETAIL)

25

20×6 FLAT TO PREVENT NUT TURNING

12 (M20)
15 (M16)

18 OR 22 DIA. HOLES

100×100×6 (43A) WASHER PLATE

WASHER PLATE DETAIL

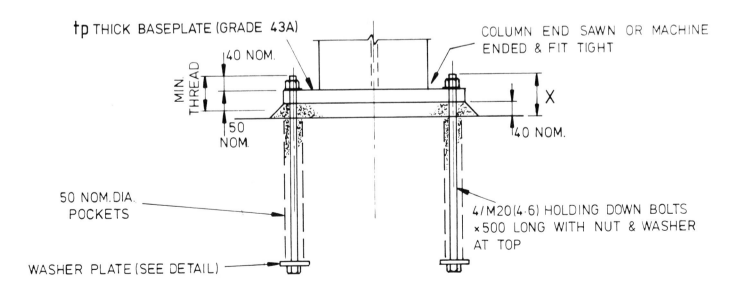

tp THICK BASEPLATE (GRADE 43A)

COLUMN END SAWN OR MACHINE ENDED & FIT TIGHT

40 NOM.

MIN. THREAD

50 NOM.

X

40 NOM.

50 NOM. DIA. POCKETS

WASHER PLATE (SEE DETAIL)

4/M20 (4·6) HOLDING DOWN BOLTS × 500 LONG WITH NUT & WASHER AT TOP

Figure 3.2 *Contd*

Table 3.5 Black bolt capacities

4.6 BOLTS IN MATERIAL GRADE 43 AND 50

Diam of Bolt mm	Tensile Stress Area mm²	Tensile Cap kN	Shear Value Single Shear kN	Shear Value Double Shear kN	Bearing Value of bolt at 435N/mm² and end distance equal to 2×bolt diameter — Thickness in mm of Plate Passed Through 5	6	7	8	9	10	12.5	15	20	25	30
12	84.3	16.4	13.5	27.0	26	*31*	*0*	*0*	*0*	*0*	*0*	*0*	*0*	*0*	*0*
16	157	30.6	25.1	50.2	34	41	48	*55*	*0*	*0*	*0*	*0*	*0*	*0*	*0*
20	245	47.8	39.2	78.4	43	52	60	69	78	*87*	*0*	*0*	*0*	*0*	*0*
22	303	59.1	48.5	97.0	**47**	57	67	76	86	95	*120*	*0*	*0*	*0*	*0*
24	353	68.8	56.5	113	**52**	62	73	83	94	104	*131*	*0*	*0*	*0*	*0*
27	459	89.5	73.4	147	**58**	**70**	82	94	106	117	147	*176*	*0*	*0*	*0*
30	561	109	89.8	180	**65**	**78**	91	104	117	131	163	*196*	*0*	*0*	*0*

8.8 BOLTS IN MATERIAL GRADE 43

Diam of Bolt mm	Tensile Stress Area mm²	Tensile Cap kN	Shear Value Single Shear kN	Shear Value Double Shear kN	Bearing Value of Plate at 460N/mm² and end distance equal to 2×bolt diameter — Thickness in mm of Plate Passed Through 5	6	7	8	9	10	12.5	15	20	25	30
12	84.3	37.9	31.6	63.2	**27**	33	38	44	49	55	*69*	*0*	*0*	*0*	*0*
16	157	70.7	58.9	118	**36**	**44**	**51**	58	66	73	93	110	*147*	*0*	*0*
20	245	110	91.9	184	**46**	**55**	**64**	**73**	**82**	92	115	138	*184*	*0*	*0*
22	303	136	114	227	**50**	**60**	**70**	**81**	**91**	**101**	127	152	202	*253*	*0*
24	353	159	132	265	**55**	**66**	**77**	**88**	**99**	**110**	138	166	221	*276*	*0*
27	459	207	172	344	**62**	**74**	**86**	**99**	**112**	**124**	**155**	186	248	310	*373*
30	561	252	210	421	**69**	**82**	**96**	**110**	**124**	**138**	**173**	**207**	276	345	414

8.8 BOLTS IN MATERIAL GRADE 50

Diam of Bolt mm	Tensile Stress Area mm²	Tensile Cap kN	Shear Value Single Shear kN	Shear Value Double Shear kN	Bearing Value of plate at 550N/mm² and end distance equal to 2×bolt diameter — Thickness in mm of Plate Passed Through 5	6	7	8	9	10	12.5	15	20	25	30
12	84.3	37.9	31.6	63.2	33	39	46	52	59	*66*	*0*	*0*	*0*	*0*	*0*
16	157	70.7	58.9	118	**44**	52	61	70	79	88	110	*132*	*0*	*0*	*0*
20	245	110	91.9	184	**55**	**66**	77	88	99	110	138	165	*220*	*0*	*0*
22	303	136	114	227	**60**	**72**	**84**	96	109	121	151	181	*242*	*0*	*0*
24	353	159	132	265	**66**	**79**	**92**	**106**	**119**	**132**	165	198	264	*330*	*0*
27	459	207	172	344	**74**	**89**	**104**	**119**	**134**	**148**	186	223	297	*377*	*0*
30	561	252	210	421	**82**	**99**	**116**	**132**	**148**	**165**	**206**	247	330	413	*495*

Values printed in bold type are less than the single shear value of the bolt. Values printed in ordinary type are greater than the single shear value and less than the double shear value. Values printed in italic type are greater than the double shear value.
Bearing values are governed by the strength of the bolt

Table 3.6 HSFG bolt capacities

IN MATERIAL GRADE 50

Diam of Bolt mm	Proof Load Of Bolt kN	Tensile Cap kN	Slip Value Single Shear kN	Slip Value Double Shear kN	Bearing Value of Plate at 1065N/mm² and end distance equal to 3xbolt diameter — Thickness in mm of Plate Passed Through										
					5	6	7	8	9	10	12.5	15	20	25	30
12	49.4	44.5	24.5	48.9	63	0	0	0	0	0	0	0	0	0	0
16	92.1	82.9	45.6	91.2	85	102	0	0	0	0	0	0	0	0	0
20	144	130	71.3	143	106	128	149	0	0	0	0	0	0	0	0
22	177	159	87.6	175	117	141	164	187	0	0	0	0	0	0	0
24	207	186	102	205	128	153	179	204	230	0	0	0	0	0	0
27	234	211	116	232	144	173	201	230	259	0	0	0	0	0	0
30	286	257	142	283	160	192	224	256	288	0	0	0	0	0	0

IN MATERIAL GRADE 43

Diam of Bolt mm	Proof Load Of Bolt kN	Tensile Cap kN	Slip Value Single Shear kN	Slip Value Double Shear kN	Bearing Value of Plate at 825N/mm² and end distance equal to 3xbolt diameter — Thickness in mm of Plate Passed Through										
					5	6	7	8	9	10	12.5	15	20	25	30
12	49.4	44.5	24.5	48.9	49	0	0	0	0	0	0	0	0	0	0
16	92.1	82.9	45.6	91.2	66	79	92	0	0	0	0	0	0	0	0
20	144	130	71.3	143	82	99	116	132	148	0	0	0	0	0	0
22	177	159	87.6	175	90	109	127	145	163	181	0	0	0	0	0
24	207	186	102	205	99	119	139	158	178	198	247	0	0	0	0
27	234	211	116	232	111	134	156	178	200	223	278	0	0	0	0
30	286	257	142	283	124	148	173	198	223	247	309	0	0	0	0

Values printed in bold type are less than the single shear value of the bolt. Values printed in ordinary type are greater than the single shear value and less than the double shear value. Values printed in italic type are greater than the double shear value.
Bearing values are governed by the strength of the plate
Slip Capacity based on a slip factor of 0.45

Table 3.7 Weld capacities

(a) Strength of fillet welds

Leg length mm	Throat thickness mm	Capacity at 215 N/mm² kN/m	Leg Length mm	Throat thickness mm	Capacity at 215 N/mm² kN/m
3.0	2.12	456	12.0	8.49	1824
4.0	2.83	608	15.0	10.61	2280
5.0	3.54	760	18.0	12.73	2737
6.0	4.24	912	20.0	14.14	3041
8.0	5.66	1216	22.0	15.56	3345
10.0	7.07	1520	25.0	17.68	3801

Capacities with grade E43 electrodes to BS 639 (22) grade of steel 43 and 50.

Table 3.7 cont'd

(b) Strength of full penetration butt welds

Thickness mm	Shear at 0.6 × Py kN/m	Tension or compression at Py kN/m	Thickness mm	Shear at 0.6 × Py kN/m	Tension or compression at Py kN/m
Grade of steel 50					
6.0	1278	2130	22.0	4554	7590
8.0	1704	2840	25.0	5175	8625
10.0	2130	3550	28.0	5796	9660
12.0	2556	4260	30.0	6210	10350
15.0	3195	5325	35.0	7245	12075
18.0	3726	6210	40.0	8280	13800
20.0	4140	6900	45.0	9180	15300
Grade of steel 43					
6.0	990	1650	22.0	3498	5830
8.0	1320	2200	25.0	3975	6625
10.0	1650	2750	28.0	4452	7420
12.0	1980	3300	30.0	4770	7950
15.0	2475	4125	35.0	5565	9275
18.0	2862	4770	40.0	6360	10600
20.0	3180	5300	45.0	6885	11475

4 DETAILING DATA

The data following provides useful information for the detailing of steelwork. The dimensional information on standard sections is given by permission of the British Steel Corporation. These sections are widely used in many other countries.

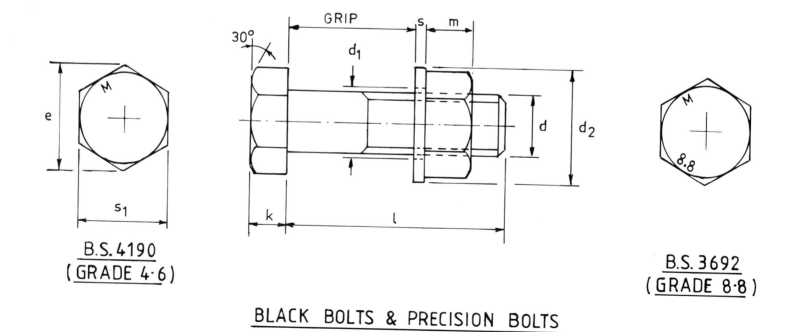

B.S. 4190
(GRADE 4·6)

B.S. 3692
(GRADE 8·8)

BLACK BOLTS & PRECISION BOLTS

Table 4.1 Dimensions of black bolts

Dimensions to BS 4190 (nearest mm)													
Bolts and Nuts				Bolts				Nuts		Washers (to BS 4320)			
Nominal Size	Pitch of Thread	Width Across Flats (max)	Width Across Corners (max)	Depth of Head (max)	Standard Length of Thread			Depth m (max)		Diameter			Thickness (Nom)
					l					Inside (Nom)	Outside (max) d2		
d		S1	e	k	≤ 125	≤ 200	> 200	Std	Thin	d1	Normal	Large	s
(M12)	1.74	19	22	9	30	36	49	11	7	14	24	28	3
M16	2.0	24	28	11	38	44	57	14	9	18	30	34	3
M20	2.5	30	35	14	46	52	65	17	9	22	37	39	3
(M22)	2.5	32	37	15	50	56	69	19	10	24	39	44	3
M24	3.0	36	42	16	54	60	73	20	10	26	44	50	4
(M27)	3.0	41	47	18	60	66	79	23	12	30	50	56	4
M30	3.5	46	53	20	66	72	85	25	12	33	56	60	4
(M33)	3.5	50	58	22	72	78	91	27	14	36	60	66	5
M36	4.0	55	64	24	78	84	97	30	14	39	66	76	5

Mechanical Properties					
		Grade 4.6 (BS 4190)		Grade 8.8 (BS 3692)	
Nominal Size	Tensile Stress Area	Ultimate Load	Proof Load	Ultimate Load	Proof Load
d	mm²	kN	kN	kN	kN
(M12)	84.3	33.1	18.7	66.2	48.1
M16	157	61.6	34.8	123	89.6
M20	245	96.1	54.3	192	140
(M22)	303	118.8	67.3	238	173
M24	353	138	78.2	277	201
(M27)	459	180	102	360	262
M30	561	220	124	439	321
(M33)	694	272	154	544	396
M36	817	321	181	641	466

Recommended Bolt and Nut Combinations										
Grade of Bolt	4.6	4.8	5.6	5.8	6.6	6.8	8.8	10.9	12.9	14.9
Recommended Grade of Nut	4	4	5	5	6	6	8	12	12	14

Notes

The single grade number for nuts indicates one tenth of the proof stress in kgf/mm² and corresponds with the bolt ultimate strength to which it is matched. It is permissible to use a higher strength grade nut than the matching bolt number. Grade 10.9 bolts are supplied with grade 12 nuts because grade 10 does not appear in the British Standard series.

Ordering example

Bolts M24 size 80mm long, grade 8.8 with standard length of thread. With standard nut grade 8.8 and normal washer. All cadmium plated.

Bolts M24 × 80 to BS 3692 – 8.8 with standard nut and normal washer. All plated to BS 3382: Part 1.

Table 4.2 Dimensions of HSFG bolts

Dimensions to BS 4395:Parts 1 & 2 (Nearest mm)																	
Bolts and Nuts						Bolts				Nuts	Washers						Add to Grip for Length
				Washer Face		Hex Head Depth (max)	Countersunk Head			Depth (max)	Round				Tapered		
Nominal Size	Pitch of Thread	Width Across Flats (max)	Width Across Girders (max)	Dia. (max)	Depth		Dia	Flash	Min Ply		Diameter		Thickness (nom)	Clip	Overall Size	Inside Dia (nom)	
											Inside (nom)	Outside (max)					
d	–	s1	e	g	b	k	j	h	p	m	d1	d2	s	w	c	d3	–
(M12)	1.75	22	25	22	0.4	9	24	2	9	12	14	30	3	12	–	–	22
M16	2.0	27	31	27	0.4	11	32	2	9	16	18	37	3	14	38	18	26
M20	2.5	32	37	32	0.4	14	40	3	12	19	21	44	4	18	38	21	30
(M22)	2.5	36	42	36	0.4	15	44	3	13	20	23	50	4	19	45	23	34
M24	3.0	41	47	41	0.5	16	48	4	15	23	26	56	4	21	57	26	36
(M27)	3.0	46	53	46	0.5	18	54	4	16	25	29	60	4	23	57	29	39
M30	3.5	50	58	50	0.5	20	60	5	19	27	33	66	4	26	57	33	42
(M33)	3.5	55	64	55	0.5	22	66	5	20	30	36	75	5	29	57	36	45
M36	4.0	60	69	60	0.5	24	72	5	22	32	–	–	–	–	–	–	48

Mechanical Properties to BS 4395:Parts 1 & 2							
		General Grade Part 1			Higher Grade Part 2		
Nominal Size	Tensile Stress Area	Proof Load	Yield Load	Ultimate Load	Proof Load	Yield Load	Ultimate Load
d	mm²	kN	kN	kN	kN	kN	kN
(M12)	84.3	49.4	53.3	69.6	–	–	–
M16	157	92.1	99.7	130	122.2	138.7	154.1
M20	245	144	155	203	190.4	216	240
(M22)	303	177	192	250	235.5	266	269.5
M24	358	207	225	292	274.6	312	345
(M27)	459	234	259	333	356	406	450
(M30)	561	286	313	406	435	495	550
(M33)	694	–	–	–	540	612	680
M36	817	418	445	591	–	–	–

Notes to Table 4.1

1. Commonly used sizes are underlined. Non-preferred sizes shown in brackets. Preferred larger diameters are M42, M56 and M64.

2. Bolt length (l) normally available in 5 mm increments up to 80 mm length and in 10 mm increments thereafter.

3. Sizes M16, M20, M24 and M27 up to 12.5 mm length may alternatively have a shorter thread length of 1½d, if so ordered. This may be required where the design does not allow the threaded portions across a shear plane.

4. BS 4190 covers black bolts of grades 4.6, 4.8 and 6.9. BS 3692 covers precision bolts in grades 4.6, 4.8, 5.6, 5.8, 6.6, 6.8, 8.8, 10.9, 12.9 and 14.9. Tolerances are closer and the maximum dimensions here quoted are slightly reduced.

Notes to Table 4.2

1. 'Add to Grip For Length' allows for nut, one flat round washer and sufficient thread protrusion beyond nut.

2. Bolt length (1) normally available in 5 mm increments up to 100 mm length and 10 mm increments thereafter. 10 mm increments are normally stocked by suppliers.

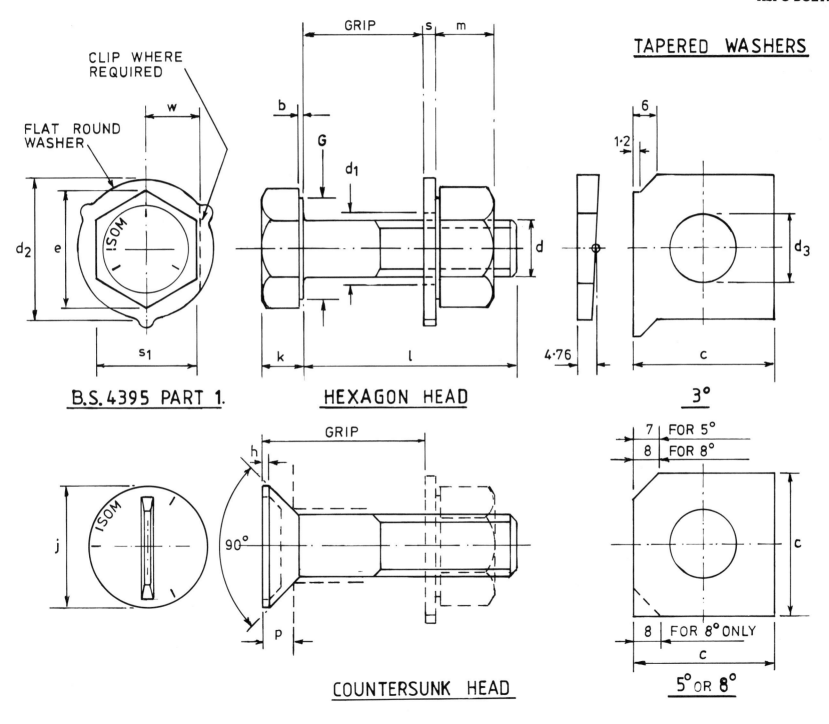

CLIP WHERE
REQUIRED

FLAT ROUND
WASHER

B.S. 4395 PART 1.

HEXAGON HEAD

TAPERED WASHERS

3°

COUNTERSUNK HEAD

5° OR 8°

Table 4.3 Universal beams   To BS 4:Part 1:1980

The Dimension $\mathbf{N} = [B - C + 6]$ to the nearest 2 mm above. $\mathbf{n} = \dfrac{D - d}{2}$ to the nearest 2mm above. $\mathbf{C} = t/2 + 2$ mm to the nearest 1 mm.

Designation		Depth	Width	Thickness		Root Radius	Web	End Clearance	Notch		Hole spacings				Max. hole dia.	Area of Section	Surface Area per metre
Serial size	Mass per metre	D	B	Web t	Flange T	r	Depth between fillets d	C	N	n	S_1	S_2	S_3	S_4			
mm	kg	mm	mm	mm	mm	mm	mm	mm	mm	mm	mm	mm	mm	mm	mm	cm²	m²
914 × 419	388	920.5	420.5	21.5	36.6	24.1	799.0	13	208	62	140	140	75	290	24	494.5	3.404
	343	911.4	418.5	19.4	32.0	24.1	799.0	12	208	58						437.5	3.382
914 × 305	289	926.6	307.8	19.6	32.0	19.1	824.4	12	152	52	140	120	60	240	20	368.8	2.988
	253	918.5	305.5	17.3	27.9	19.1	824.4	11	152	48						322.8	2.967
	224	910.3	304.1	15.9	23.9	19.1	824.4	10	154	44						285.3	2.948
	201	903.0	303.4	15.2	20.2	19.1	824.4	10	152	40						256.4	2.932
838 × 292	226	850.9	293.8	16.1	26.8	17.8	761.7	10	148	46	140	–	–	–	24	288.7	2.791
	194	840.7	292.4	14.7	21.7	17.8	761.7	9	148	40						247.2	2.767
	176	834.9	291.6	14.0	18.8	17.8	761.7	9	148	38						224.1	2.754
762 × 267	197	769.6	268.0	15.6	25.4	16.5	685.8	10	134	42	140	–	–	–	24	250.8	2.530
	173	762.0	266.7	14.3	21.6	16.5	685.8	9	136	40						220.5	2.512
	147	753.9	265.3	12.9	17.5	16.5	685.8	8	136	36						188.1	2.493
686 × 254	170	692.9	255.8	14.5	23.7	15.2	615.0	9	130	40	140	–	–	–	24	216.6	2.333
	152	687.6	254.5	13.2	21.0	15.2	615.0	9	130	38						193.8	2.320
	140	683.5	253.7	12.4	19.0	15.2	615.0	8	130	36						178.6	2.310
	125	677.9	253.0	11.7	16.2	15.2	615.0	8	130	32						159.6	2.298
610 × 305	238	633.0	311.5	18.6	31.4	16.5	537.2	11	156	48	140	120	60	240	20	303.8	2.421
	179	617.5	307.0	14.1	23.6	16.5	537.2	9	156	42						227.9	2.381
	149	609.6	304.8	11.9	19.7	16.5	537.2	8	156	38						190.1	2.361
610 × 229	140	617.0	230.1	13.1	22.1	12.7	547.2	9	118	36	140	–	–	–	24	178.4	2.088
	125	611.9	229.0	11.9	19.6	12.7	547.2	8	118	34						159.6	2.075
	113	607.3	228.2	11.2	17.3	12.7	547.2	8	118	32						144.5	2.064
	101	602.2	227.6	10.6	14.8	12.7	547.2	7	118	28						129.2	2.053
533 × 210	122	544.6	211.9	12.8	21.3	12.7	476.5	8	108	34	140	–	–	–	24	155.8	1.872
	109	539.5	210.7	11.6	18.8	12.7	476.5	8	108	32						138.6	1.860
	101	536.7	210.1	10.9	17.4	12.7	476.5	7	110	30						129.3	1.853
	92	533.1	209.3	10.2	15.6	12.7	476.5	7	108	30						117.8	1.844
	82	528.3	208.7	9.6	13.2	12.7	476.5	7	108	26						104.4	1.833
457 × 191	98	467.4	192.8	11.4	19.6	10.2	407.9	8	100	30	90	–	–	–	24	125.3	1.650
	89	463.6	192.0	10.6	17.7	10.2	407.9	7	100	28						113.9	1.641
	82	460.2	191.3	9.9	16.0	10.2	407.9	7	100	28						104.5	1.633
	74	457.2	190.5	9.1	14.5	10.2	407.9	7	100	26						95.0	1.625
	67	453.6	189.9	8.5	12.7	10.2	407.9	6	100	24						85.4	1.617

Table 4.3 *Contd*

| Designation | | Depth | Width | Thickness | | Root Radius | Web | End Clearance | Notch | | Hole spacing | Max. hole dia. | Area of Section | Surface Area |
| Serial size | Mass per metre | D | B | Web t | Flange T | r | Depth between fillets D | C | N | n | S₁ | | | per metre |
mm	kg	mm	mm	mm	mm	mm	mm	mm	mm	mm	mm	mm	cm²	m²
457 × 152	82	465.1	153.5	10.7	18.9	10.2	406.9	7	80	30			104.5	1.493
	74	461.3	152.7	9.9	17.0	10.2	406.9	7	80	28			95.0	1.484
	67	457.2	151.9	9.1	15.0	10.2	406.9	7	80	26	90	20	85.4	1.474
	60	454.7	152.9	8.0	13.3	10.2	407.7	6	82	24			75.9	1.487
	52	449.8	152.4	7.6	10.9	10.2	407.7	6	82	22			66.5	1.476
406 × 178	74	412.8	179.7	9.7	16.0	10.2	360.5	7	94	28			95.0	1.493
	67	409.4	178.8	8.8	14.3	10.2	360.5	6	94	26	90	24	85.5	1.484
	60	406.4	177.8	7.8	12.8	10.2	360.5	6	94	24			76.0	1.476
	54	402.6	177.6	7.6	10.9	10.2	360.5	6	94	22			68.4	1.468
406 × 140	46	402.3	142.4	6.9	11.2	10.2	359.6	5	78	24	70	20	59.0	1.332
	39	397.3	141.8	6.3	8.6	10.2	359.6	5	76	20			49.4	1.320
356 × 171	67	364.0	173.2	9.1	15.7	10.2	312.2	7	90	26			85.4	1.371
	57	358.6	172.1	8.0	13.0	10.2	312.2	6	92	24	90	24	72.2	1.358
	51	355.6	171.5	7.3	11.5	10.2	312.2	6	90	22			64.6	1.351
	45	352.0	171.0	6.9	9.7	10.2	312.2	5	92	20			57.0	1.343
356 × 127	39	352.8	126.0	6.5	10.7	10.2	311.1	5	68	22	70	20	49.4	1.169
	33	348.5	125.4	5.9	8.5	10.2	311.1	5	68	20			41.8	1.160
305 × 165	54	310.9	166.8	7.7	13.7	8.9	265.6	6	88	24			68.4	1.245
	46	307.1	165.7	6.7	11.8	8.9	265.6	5	88	22	90	24	58.9	1.235
	40	303.8	165.1	6.1	10.2	8.9	265.6	5	88	20			51.5	1.227
305 × 127	48	310.4	125.2	8.9	14.0	8.9	264.6	6	68	24			60.8	1.079
	42	306.6	124.3	8.0	12.1	8.9	264.6	6	68	22	70	20	53.2	1.069
	37	303.8	123.5	7.2	10.7	8.9	264.6	6	66	20			47.5	1.062
305 × 102	33	312.7	102.4	6.6	10.8	7.6	275.8	5	58	20			41.8	1.006
	28	308.9	101.9	6.1	8.9	7.6	275.8	5	56	18	54	12	36.3	0.997
	25	304.8	101.6	5.8	6.8	7.6	275.8	5	56	16			31.4	0.988
254 × 146	43	259.6	147.3	7.3	12.7	7.6	218.9	6	78	22			55.1	1.069
	37	256.0	146.4	6.4	10.9	7.6	218.9	5	80	20	70	20	47.5	1.060
	31	251.5	146.1	6.1	8.6	7.6	218.9	5	80	18			40.0	1.050
254 × 102	28	260.4	102.1	6.4	10.0	7.6	225.0	5	58	18			36.2	0.900
	25	257.0	101.9	6.1	8.4	7.6	225.0	5	56	16	54	12	32.2	0.893
	22	254.0	101.6	5.8	6.8	7.6	225.0	5	56	16			28.4	0.887
203 × 133	30	206.8	133.8	6.3	9.6	7.6	172.3	5	72	18	70	20	38.0	0.912
	25	203.2	133.4	5.8	7.8	7.6	172.3	5	72	16			32.3	0.904

Table 4.3 *Contd*

Designation		Depth	Width	Thickness		Root Radius	Web	End Clearance	Notch		Hole spacing	Max. hole dia.	Area of Section	Surface Area
Serial size	Mass per metre	D	B	Web t	Flange T	r	Depth between fillets D	C	N	n	S₁			per metre
mm	kg	mm	mm	mm	mm	mm	mm	mm	mm	mm	**mm**	mm	cm²	m²
203 × 102	23	203.2	101.6	5.2	9.3	7.6	169.4	5	56	18	**54**	12	29.0	0.789
178 × 102	19	177.8	101.6	4.7	7.9	7.6	146.8	4	58	16	**54**	12	24.2	0.740
152 × 89	16	152.4	88.9	4.6	7.7	7.6	121.8	4	52	16	**50**	–	20.5	0.638
127 × 76	13	127.0	76.2	4.2	7.6	7.6	96.6	4	46	16	**40**	–	16.8	0.537

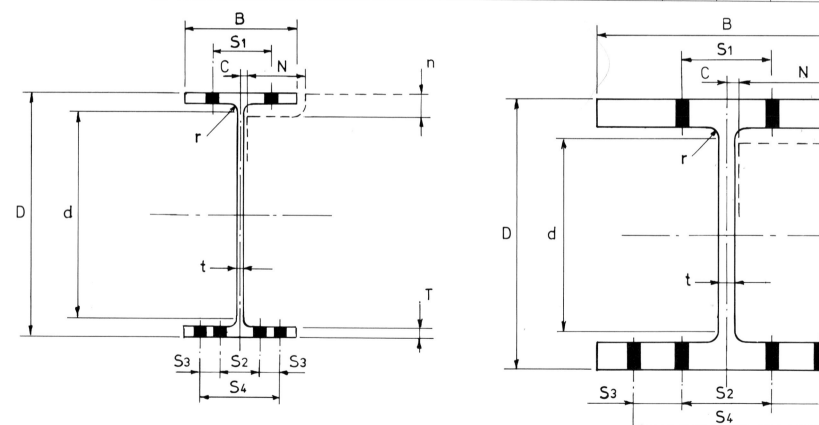

See Table 4.3 UNIVERSAL BEAMS

See Table 4.4 UNIVERSAL COLUMNS

Table 4.4 Universal columns

The Dimension $N = [B - C + 6]$ to the nearest 2 mm above.

$n = \dfrac{D - d}{2}$ to the nearest 2mm above.

$C = t/2 + 2$ mm to the nearest 1 mm.

To BS 4:Part 1:1980

Designation Serial size	Mass per metre	Depth of Section D	Width of Section B	Thickness Web t	Thickness Flange T	Root Radius r	Web Depth between fillets d	End Clearance C	Notch N	Notch n	S_1	S_2	S_3	S_4	Max. hole dia.	Area of Section	Surface Area per metre
mm	kg	mm	mm	mm	mm	mm	mm	mm	mm	mm	mm	mm	mm	mm	mm	cm²	m²
356 × 406	634	474.7	424.1	47.6	77.0	15.2	290.1	26	198	94						808.1	2.525
	551	455.7	418.5	42.0	67.5	15.2	290.1	23	198	84						701.8	2.475
	467	436.6	412.4	35.9	58.0	15.2	290.1	20	198	74						595.5	2.425
	393	419.1	407.0	30.6	49.2	15.2	290.1	17	198	66	140	140	75	290	24	500.9	2.379
	340	406.4	403.0	26.5	42.9	15.2	290.1	15	198	60						432.7	2.346
	287	393.7	399.0	22.6	36.5	15.2	290.1	13	198	52						366.0	2.312
	235	381.0	395.0	18.5	30.2	15.2	290.1	11	198	46						299.8	2.279
356 × 368	202	374.7	374.4	16.8	27.0	15.2	290.1	10	188	44						257.9	2.187
	177	368.3	372.1	14.5	23.8	15.2	290.1	9	188	44	140	140	75	290	24	225.7	2.170
	153	362.0	370.2	12.6	20.7	15.2	290.1	8	188	36						195.2	2.154
	129	355.6	368.3	10.7	17.5	15.2	290.1	7	188	34						164.9	2.137
305 × 305	283	365.3	321.3	26.9	44.1	15.2	246.6	15	156	60						360.4	1.938
	240	352.6	317.9	23.0	37.7	15.2	246.6	13	156	54	140	120	60	240	24	305.6	1.905
	198	339.9	314.1	19.2	31.4	15.2	246.6	12	156	48						252.3	1.872
	158	327.2	310.6	15.7	25.0	15.2	246.6	10	156	42						201.2	1.839
	137	320.5	308.7	13.8	21.7	15.2	246.6	9	156	38	140	120	60	240	20	174.6	1.822
	118	314.5	306.8	11.9	18.7	15.2	246.6	8	156	34						149.8	1.806
	97	307.8	304.8	9.9	15.4	15.2	246.6	7	156	32						123.3	1.789
254 × 254	167	289.1	264.5	19.2	31.7	12.7	200.2	12	132	46						212.4	1.576
	132	276.4	261.0	15.6	25.3	12.7	200.2	10	132	40						168.9	1.543
	107	266.7	258.3	13.0	20.5	12.7	200.2	9	132	34	140	–	–	–	24	136.6	1.519
	89	260.4	255.9	10.5	17.3	12.7	200.2	7	132	32						114.0	1.502
	73	254.0	254.0	8.6	14.2	12.7	200.2	6	132	28						92.9	1.485
203 × 203	86	222.3	208.8	13.0	20.5	10.2	160.8	8	108	32						110.1	1.236
	71	215.9	206.2	10.3	17.3	10.2	160.8	7	108	28						91.1	1.218
	60	209.6	205.2	9.3	14.2	10.2	160.8	7	106	26	140	–	–	–	24	75.8	1.204
	52	206.2	203.9	8.0	12.5	10.2	160.8	6	106	24						66.4	1.194
	46	203.2	203.2	7.3	11.0	10.2	160.8	6	106	22						58.8	1.187
152 × 152	37	161.8	154.4	8.1	11.5	7.6	123.4	6	82	20						47.4	0.912
	30	157.5	152.9	6.6	9.4	7.6	123.4	5	82	18	90	–	–	–	20	38.2	0.900
	23	152.4	152.4	6.1	6.8	7.6	123.4	5	82	16						29.8	0.889

Table 4.5 Joists

The Dimension $N = [B - C + 6]$ to the nearest 2 mm above.

$n = \dfrac{D - d}{2}$ to the nearest 2mm above.

$C = [t/2 + 2]$ to the nearest 1 mm.

To BS 4:Part 1:1980

Designation		Depth of Section D	Width of Section B	Thickness		Root radius r_1	Toe radius r_2	Inside Slope	Depth d	End Clearance C	Notch		S	Max. hole dia.	Area of Section	Surface Area per metre
Serial size	Mass per metre			Web t	Flange T						N	n				
mm	kg	mm	mm	mm	mm	mm	**mm**	degrees	mm	mm	mm	mm	mm	mm	cm²	m²
254 × 203	81.85	254.0	203.2	10.2	19.9	19.6	**9.7**	8	166.6	7	106	45	140	24	104.4	1.193
254 × 114	37.20	254.0	114.3	7.6	12.8	12.4	**6.1**	8	199.1	6	62	30	65	16	47.4	0.882
203 × 152	52.09	203.2	152.4	8.9	16.5	15.5	**7.6**	8	133.4	7	80	40	90	20	66.4	0.911
152 × 127	37.20	152.4	127.0	10.4	13.2	13.5	**6.6**	8	94.5	7	68	35	70	20	47.5	0.722
127 × 114	29.76	127.0	114.3	10.2	11.5	12.4	**4.8**	8	71.9	7	62	30	65	16	37.3	0.620
127 × 114	26.79	127.0	114.3	7.4	11.4	9.9	**5.0**	8	79.2	6	62	25	65	16	34.1	0.635
127 × 76	16.37	127.0	76.2	5.6	9.6	9.4	**4.6**	8	86.4	5	44	25	40	–	21.0	0.498
114 × 114	26.79	114.3	114.3	9.5	10.7	14.2	**3.2**	8	61.1	7	62	30	65	16	34.4	0.600
102 × 102	23.07	101.6	101.6	9.5	10.3	11.1	**3.2**	8	54.0	7	54	25	54	12	29.4	0.528
89 × 89	19.35	88.9	88.9	9.5	9.9	11.1	**3.2**	8	45.2	7	48	25	50	–	24.9	0.460
76 × 76	12.65	76.2	76.2	5.1	8.4	9.4	**4.6**	8	38.1	5	40	20	40	–	16.3	0.403

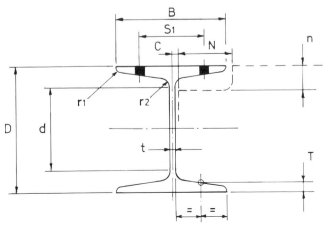

Table 4.6 Channels

The Dimension $N = [B - C + 6]$ to the nearest 2 mm above. $n = \dfrac{D - d}{2}$ to the nearest 2mm above. $C = [t + 2]$ to the nearest 1 mm.

To BS 4:Part 1:1980

Designation		Depth	Width	Thickness		Distance of	Root radius	Toe radius	Inside Slope	Depth		Notch			Max. hole dia.	Area of Section	Surface Area per metre
Serial size	Mass per metre	D	B	Web t	Flange T	y	r_1	r_2		d	C	N	n	S			
mm	kg	mm	mm	mm	mm	cm	mm	mm	degrees	mm	mm	mm	mm	mm	mm	cm²	m²
432 × 102	65.54	431.8	101.6	12.2	16.8	2.32	15.2	4.8	5	362.5	14	94	36	55	20	83.49	1.21
381 × 102	55.10	381.0	101.6	10.4	16.3	2.52	15.2	4.8	5	312.6	12	96	36	55	20	70.19	1.11
305 × 102	46.18	304.8	101.6	10.2	14.8	2.66	15.2	4.8	5	239.3	12	96	34	55	20	58.83	0.96
305 × 89	41.69	304.8	88.9	10.2	13.7	2.18	13.7	3.2	5	245.4	12	84	30	50	20	53.11	0.92
254 × 89	35.74	254.0	88.9	9.1	13.6	2.42	13.7	3.2	5	194.7	11	84	30	50	20	45.52	0.82
254 × 76	28.29	254.0	76.2	8.1	10.9	1.86	12.2	3.2	5	203.9	10	74	26	45	20	36.03	0.774
229 × 89	32.76	228.6	88.9	8.6	13.3	2.53	13.7	3.2	5	169.9	11	84	30	50	20	41.73	0.770
229 × 76	26.06	228.6	76.2	7.6	11.2	2.00	12.2	3.2	5	177.8	10	74	26	45	20	33.20	0.725
203 × 89	29.78	203.2	88.9	8.1	12.9	2.65	13.7	3.2	5	145.2	10	86	30	50	20	37.94	0.720
203 × 76	23.82	203.2	76.2	7.1	11.2	2.13	12.2	3.2	5	152.4	9	74	26	45	20	30.34	0.675
178 × 89	26.81	177.8	88.9	7.6	12.3	2.76	13.7	3.2	5	121.0	10	86	30	50	20	34.15	0.671
178 × 76	20.84	177.8	76.2	6.6	10.3	2.20	12.2	3.2	5	128.8	9	74	26	45	20	26.54	0.625
152 × 89	23.84	152.4	88.9	7.1	11.6	2.86	13.7	3.2	5	96.9	9	86	28	50	20	30.36	0.621
152 × 76	17.88	152.4	76.2	6.4	9.0	2.21	12.2	2.4	5	105.9	8	76	24	45	20	22.77	0.575
127 × 64	14.90	127.0	63.5	6.4	9.2	1.94	10.7	2.4	5	84.0	8	62	22	35	16	18.98	0.476
102 × 51	10.42	101.6	50.8	6.1	7.6	1.51	9.1	2.4	5	65.8	8	50	18	30	10	13.28	0.379

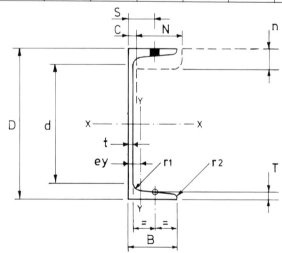

Table 4.7 Rolled steel angles: (a) equal, (see p. 75)

Note: *Not included in BS 4848:Part 4:1972

To BS 4848:Part 4:1972

Designation		Mass per metre	Area of section	Distance of centre of gravity ex, ey	Recommended back marks			Max. dia. bolt
Size B	Thickness t				S₁	S₂	S₃	
mm	mm	kg	cm²	cm	mm	mm	mm	mm
250 × 250	35	128.0	163.0	7.49				
	32	118.0	150.0	7.38	–	90	100	36
	28	104.0	133.0	7.23				
	25	93.6	119.0	7.12				
200 × 200	24	71.1	90.6	5.84				
	20	59.9	76.3	5.68	–	75	75	30
	18	54.2	69.1	5.60				
	16	48.5	61.8	5.52				
150 × 150	18	40.1	51.0	4.37				
	15	33.8	43.0	4.25	–	55	55	20
	12	27.3	34.8	4.12				
	10	23.0	29.3	4.03				
120 × 120	15	26.6	33.9	3.51				
	12	21.6	27.5	3.40	–	45	50	16
	10	18.2	23.2	3.31				
	8	14.7	18.7	3.23				
100 × 100	15	21.9	27.9	3.02				
	12	17.8	22.7	2.90	55	–	–	24
	10*	15.0	19.2	2.82				
	8	12.2	15.5	2.74				
90 × 90	12	15.9	20.3	2.66				
	10	13.4	17.1	2.58	50	–	–	24
	8	10.9	13.9	2.50				
	6	8.30	10.6	2.41				

Table 4.7 Rolled steel angles: (b) unequal, (see p. 75)

To BS 4848:Part4:1972

Designation		Mass per metre	Area of section	Distance of centre of gravity		Recommended back marks			Max. dia. bolt	
Size D × B	Thickness t			ex	ey	S_1	S_2	S_3	For S_1	For $S_2 S_3$
mm	mm	kg	cm^2	cm	cm	mm	mm	mm	mm	mm
200 × 150	18	47.1	60.0	6.33	3.85					
	15	39.6	50.5	6.21	3.73	–	75	75	–	30
	12	32.0	40.8	6.08	3.61					
200 × 100	15	33.7	43.0	7.16	2.22					
	12	27.3	34.8	7.03	2.10	55	75	75	24	30
	10	23.0	29.2	6.93	2.01					
150 × 90	15	26.6	33.9	5.21	2.23					
	12	21.6	27.5	5..08	2.12	50	55	55	24	20
	10	18.2	23.2	5.00	2.04					
150 × 75	15	24.8	31.6	5.53	1.81					
	12	20.2	25.7	5.41	1.69	45	55	55	20	20
	10	17.0	21.6	5.32	1.61					
137 × 102	9.5	17.3	21.9	4.22	2.49					
	7.9	14.5	18.4	4.17	2.44	55	50	50	24	20
	6.4	11.7	14.8	4.09	2.36					
125 × 75	12	17.8	22.7	4.31	1.84					
	10	15.0	19.1	4.23	1.76	45	45	50	20	20
	8	12.2	15.5	4.14	1.68					
	6.5	9.98	12.7	4.06	1.61					
100 × 75	12	15.4	19.7	3.27	2.03					
	10	13.0	16.6	3.19	1.95	55	–	–	24	–
	8	10.6	13.5	3.10	1.87					
100 × 65	10	12.3	15.6	3.36	1.63					
	8	9.94	12.7	3.27	1.55	55	–	–	24	–
	7	8.77	11.2	3.23	1.51					

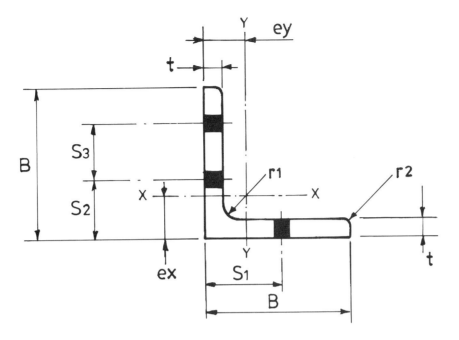

Equal angles (a)

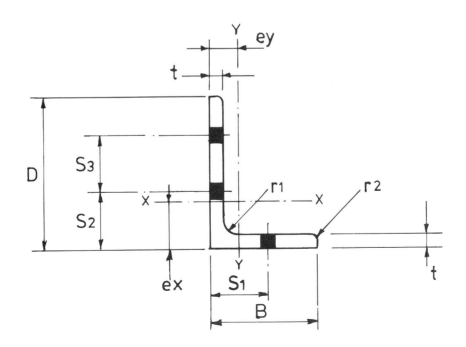

Unequal angles (b)

Table 4.8 Square hollow sections

Size D × D	Thickness t	Mass per metre M	Area of section A	Surface area per metre	Size D × D	Thickness t	Mass per metre M	Area of section A	Surface area per metre
mm	mm	kg	cm²	m²	mm	mm	kg	cm²	m²
20 x 20	2.0	1.12	1.42	0.076	120 x 120	5.0	18.0	22.9	0.469
	2.5*	1.35	1.72	0.075		6.3	22.3	28.5	0.466
25 x 25*	2.0*	1.43	1.82	0.096		8.0	27.9	35.5	0.463
	2.5*	1.74	2.22	0.095		10.0	34.2	43.5	0.459
	3.2*	2.15	2.74	0.093	140 x 140*	5.0*	21.1	26.9	0.549
30 x 30	2.5*	2.14	2.72	0.115		6.3*	26.3	33.5	0.546
	3.0*	2.51	3.20	0.114		8.0*	32.9	41.9	0.543
	3.2	2.65	3.38	0.113		10.0*	40.4	51.5	0.539
40 x 40	2.5*	2.92	3.72	0.155	150 x 150	5.0	22.7	28.9	0.589
	3.0*	3.45	4.40	0.154		6.3	28.3	36.0	0.586
	3.2	3.66	4.66	0.153		8.0	35.4	45.1	0.583
	4.0	4.46	5.68	0.151		10.0	43.6	55.5	0.579
50 x 50	2.5*	3.71	4.72	0.195		12.5*	53.4	68.0	0.573
	3.0*	4.39	5.60	0.194		16.0*	66.4	84.5	0.566
	3.2	4.66	5.94	0.193	180 x 180	6.3	34.2	43.6	0.706
	4.0	5.72	7.28	0.191		8.0	43.0	54.7	0.703
	5.0	6.97	8.88	0.189		10.0	53.0	67.5	0.699
60 x 60	3.0*	5.34	6.80	0.234		12.5*	65.2	83.0	0.693
	3.2	5.67	7.22	0.233		16.0*	81.4	104	0.686
	4.0	6.97	8.88	0.231	200 x 200	6.3	38.2	48.6	0.786
	5.0	8.54	10.9	0.229		8.0	48.0	61.1	0.783
70 x 70	3.0*	6.28	8.00	0.274		10.0	59.3	75.5	0.779
	3.6	7.46	9.50	0.272		12.5*	73.0	93.0	0.773
	5.0	10.1	12.9	0.269		16.0*	91.5	117	0.766
80 x 80	3.0*	7.22	9.20	0.314	250 x 250	6.3	48.1	61.2	0.986
	3.6	8.59	10.9	0.312		8.0	60.5	77.1	0.983
	5.0	11.7	14.9	0.309		10.0	75.0	95.5	0.979
	6.3	14.4	18.4	0.306		12.5*	92.6	118	0.973
90 x 90	3.6	9.72	12.4	0.352		16.0*	117	149	0.966
	5.0	13.3	16.9	0.349	300 x 300	10.0	90.7	116	1.18
	6.3	16.4	20.9	0.346		12.5*	112	143	1.17
100 x 100	4.0	12.0	15.3	0.391		16.0*	142	181	1.17
	5.0	14.8	18.9	0.389	350 x 350	10.0	106	136	1.38
	6.3	18.4	23.4	0.386		12.5*	132	168	1.37
	8.0	22.9	29.1	0.383		16.0*	167	213	1.37
	10.0	27.9	35.5	0.379	400 x 400	10.0	122	156	1.58
						12.5*	152	193	1.57

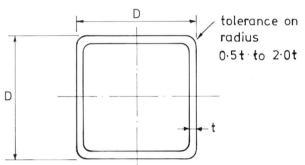

tolerance on radius
0·5t to 2·0t

Table 4.9 Rectangular hollow sections

Size D x B (mm)	Thickness t (mm)	Mass per metre M (kg)	Area of section A (cm²)	Surface area per metre (m²)	Size D x B (mm)	Thickness t (mm)	Mass per metre M (kg)	Area of section A (cm²)	Surface area per metre (m²)
50 x 25*	2.5*	2.72	3.47	0.145	150 x 100	5.0	18.7	23.9	0.489
	3.0*	3.22	4.10	0.144		6.3	23.3	29.7	0.486
	3.2*	3.41	4.34	0.143		8.0	29.1	37.1	0.483
50 x 30	2.5*	2.92	3.72	0.155		10.0	35.7	45.5	0.479
	3.0*	3.45	4.40	0.154	160 x 80	5.0	18.0	22.9	0.469
	3.2	3.66	4.66	0.153		6.3	22.3	28.5	0.466
60 x 40	2.5*	3.71	4.72	0.195		8.0	27.9	35.5	0.463
	3.0*	4.39	5.60	0.194		10.0	34.2	43.5	0.459
	3.2	4.66	5.94	0.193	200 x 100	5.0	22.7	28.9	0.589
	4.0	5.72	7.28	0.191		6.3	28.3	36.0	0.586
80 x 40	3.0*	5.34	6.80	0.234		8.0	35.4	45.1	0.583
	3.2	5.67	7.22	0.233		10.0-	43.6	55.5	0.579
	4.0	6.97	8.88	0.231		12.5	53.4	68.0	0.573
90 x 50	3.0*	6.28	8.00	0.274		16.0	66.4	84.5	0.566
	3.6	7.46	9.50	0.272	250 x 150	6.3	38.2	48.6	0.785
	5.0	10.1	12.9	0.269		8.0	48.0	61.1	0.783
100 x 50	3.0*	6.75	8.60	0.294		10.0	59.3	75.5	0.779
	3.2	7.18	9.14	0.293		12.5	73.0	93.0	0.773
	4.0	8.86	11.3	0.291		16.0	91.5	117	0.766
	5.0	10.9	13.9	0.289	300 x 200	6.3	48.1	61.2	0.986
	6.3*	13.4	17.1	0.286		8.0	60.5	77.1	0.983
100 x 60	3.0*	7.22	9.20	0.314		10.0	75.0	95.5	0.979
	3.6	8.59	10.9	0.312		12.5	92.6	118	0'973
	5.0	11.7	14.9	0.309		16.0	117	149	0.966
	6.3	14.4	18.4	0.306	400 x 200	10.0	90.7	116	1.18
120 x 60	3.6	9.72	12.4	0.352		12.5	112	143	1.17
	5.0	13.3	16.9	0.349		16.0	142	181	1.17
	6.3	16.4	20.9	0.346	450 x 250	10.0	106	136	1.38
120 x 80	5.0	14.8	18.9	0.389		12.5	132	168	1.37
	6.3	18.4	23.4	0.386		16.0	167	213	1.37
	8.0	22.9	29.1	0.383					
	10.0	27.9	35.5	0.379					

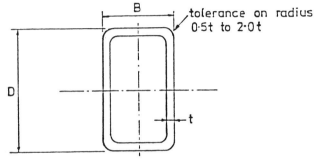

tolerance on radius 0.5t to 2.0t

Table 4.10 Circular hollow sections

* Thickness not included in BS 4848: Part 2.

Designation		Mass per metre M	Area of section A	Surface area per metre
Outside diameter D	Thickness t			
mm	mm	kg	cm²	m²
21.3	3.2	1.43	1.82	0.067
26.9	3.2	1.87	2.38	0.085
33.7	2.6	1.99	2.54	0.106
	3.2	2.41	3.07	0.106
	4.0	2.93	3.73	0.106
42.4	2.6	2.55	3.25	0.133
	3.2	3.09	3.94	0.133
	4.0	3.79	4.83	0.133
48.3	3.2	3.56	4.53	0.152
	4.0	4.37	5.57	0.152
	5.0	5.34	6.80	0.152
60.3	3.2	4.51	5.74	0.189
	4.0	5.55	7.07	0.189
	5.0	6.82	8.69	0.189
76.1	3.2	5.75	7.33	0.239
	4.0	7.11	9.06	0.239
	5.0	8.77	11.2	0.239
88.9	3.2	6.76	8.62	0.279
	4.0	8.38	10.7	0.279
	5.0	10.3	13.2	0.279
114.3	3.6	9.83	12.5	0.359
	5.0	13.5	17.2	0.359
	6.3	16.8	21.4	0.359
139.7	5.0	16.6	21.2	0.439
	6.3	20.7	26.4	0.439
	8.0	26.0	33.1	0.439
	10.0	32.0	40.7	0.439
168.3	5.0	20.1	25.7	0.529
	6.3	25.2	32.1	0.529
	8.0	31.6	40.3	0.529
	10.0	39.0	49.7	0.529

Designation		Mass per metre M	Area of section A	Surface area per metre
Outside diameter D	Thickness t			
mm	mm	kg	cm²	m²
193.7	5.0*	23.3	29.6	0.609
	5.4	25.1	31.9	0.609
	6.3	29.1	37.1	0.609
	8.0	36.6	46.7	0.609
	10.0	45.3	57.7	0.609
	12.5	55.9	71.2	0.609
	16.0	70.1	89.3	0.609
219.1	5.0*	26.4	33.6	0.688
	6.3	33.1	42.1	0.688
	8.0	41.6	53.1	0:688
	10.0	51.6	65.7	0.688
	12.5	63.7	81.1	0.688
	16.0	80.1	102	0.688
	20.0	98.2	125	0.688
244.5	6.3	37.0	47.1	0.768
	8.0	46.7	59.4	0.768
	10.0	57.8	73.7	0.768
	12.5	71.5	91.1	0.768
	16.0	90.2	115	0.768
	20.0	111	141	0.768
273	6.3	41.4	52.8	0.858
	8.0	52.3	66.6	0.858
	10.0	64.9	82.6	0.858
	12.5	80.3	102	0.858
	16.0	101	129	0.858
	20.0	125	159	0.858
	25.0	153	195	0.858
323.9	6.3*	49.3	62.9	1.02
	8.0	62.3	79.4	1.02
	10.0	77.4	98.6	1.02
	12.5	96.0	122	1.02
	16.0	121	155	1.02
	20.0	150	191	1.02
	25.0	184	235	1.02

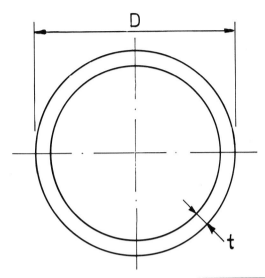

Designation				Surface
		Mass	Area	area
		per	of	per
Outside		metre	section	metre
diameter	Thickness	M	A	
D	t			
mm	mm	kg	cm²	m²
355.6	8.0	68.6	87.4	1.12
	10.0	85.2	109	1.12
	12.5	106	135	1.12
	16.0	134	171	1.12
	20.0	166	211	1.12
	25.0	204	260	1.12
406.4	10.0	97.8	125	1.28
	12.5	121	155	1.28
	16.0	154	196	1.28
	20.0	191	243	1.28
	25.0	235	300	1.28
	32.0	295	376	1.28
457	10.0	110	140	1.44
	12.5	137	175	1.44
	16.0	174	222	1.44
	20.0	216	275	1.44
	25.0	266	339	1.44
	32.0	335	427	1.44
	40.0	411	524	1.44
508	10.0*	123	156	1.60
	12.5*	153	195	1.60

Table 4.11 Metric bulb flats

Designation	Depth	Thickness	Bulb Height	Bulb Radius	Area of section	Mass per metre	Surface Area	Centroid
	b	t	c	r1	A			ex
Size mm	mm	mm	mm	mm	cm²	kg/m	m²/m	cm
120 x 6	120	6	17	5	9.31	7.31	0.276	7.20
7		7	17	5	10.5	8.25	0.278	7.07
8		8	17	5	11.7	9.19	0.280	6.96
140 x 6.5	140	6.5	19	5.5	11.7	9.21	0.319	8.37
7		7	19	5.5	12.4	9.74	0.320	8.31
8		8	19	5.5	13.8	10.83	0.322	8.18
10		10	19	5.5	16.6	13.03	0.326	7.92
160 x 7	160	7	22	6	14.6	11.4	0.365	9.66
8		8	22	6	16.2	12.7	0.367	9.49
9		9	22	6	17.8	14.0	0.369	9.36
11.5		11.5	22	6	21.8	17.3	0.374	9.11
180 x 8	180	8	25	7	18.9	14.8	0.411	10.9
9		9	25	7	20.7	16.2	0.413	10.7
10		10	25	7	22.5	17.6	0.415	10.6
11.5		11.5	25	7	25.2	19.7	0.418	10.4
200 x 8.5	200	8.5	28	8	22.6	17.8	0.456	12.2
9		9	28	8	23.6	18.5	0.457	12.1
10		10	28	8	25.6	20.1	0.459	11.9
11		11	28	8	27.6	21.7	0.461	11.8
12		12	28	8	29.6	23.2	0.463	11.7
220 x 9	220	9	31	9	26.8	21.0	0.501	13.6
10		10	31	9	29.0	22.8	0.503	13.4
11		11	31	9	31.2	24.5	0.505	13.2
12		12	31	9	33.4	26.2	0.507	13.0
240 x 9.5	240	9.5	34	10	31.2	24.4	0.546	14.8
10		10	34	10	32.4	25.4	0.547	14.7
11		11	34	10	34.9	27.4	0.549	14.6
12		12	34	10	37.3	29.3	0.551	14.4
260 x 10	260	10	37	11	36.1	28.3	0.593	16.2
11		11	37	11	38.7	30.3	0.593	16.0
12		12	37	11	41.3	32.4	0.595	15.8

Table 4.11 *Contd*

Designation	Depth	Thickness	Bulb Height	Bulb Radius	Area of section	Mass per metre	Surface Area	Centroid	Moment of Inertia	Modulus
	D	t	c	r1	A			ex	Ixx	Zxx
Size mm	mm	mm	mm	mm	cm²	kg/m	m²/m	cm	cm⁴	cm³
280 x 10.5	280	10.5	40	12	41.2	32.4	0.636	17.5	3223	184
11		11	40	12	42.6	33.5	0.637	17.4	3330	191
12		12	40	12	45.5	35.7	0.639	17.2	3550	206
13		13	40	12	48.4	37.9	0.641	17.0	3760	221
300 x 11	300	11	43	13	46.7	36.7	0.681	18.9	4190	222
12		12	43	13	49.7	39.0	0.683	18.7	4460	239
13		13	43	13	52.8	41.5	0.685	18.5	4720	256
320 x 11.5	320	11.5	46	14	52.6	41.2	0.727	20.2	5370	266
12		12	46	14	54.2	42.5	0.728	20.1	5530	274
13		13	46	14	57.4	45.0	0.730	19.9	5850	294
14		14	46	14	60.6	47.5	0.732	19.7	6170	313
340 x 12	340	12	49	15	58.8	46.1	0.772	21.5	6760	313
13		13	49	15	62.2	48.8	0.774	21.3	7160	335
14		14	49	15	65.5	51.5	0.776	21.1	7540	357
15		15	49	15	69.0	54.2	0.778	20.9	7920	379
370 x 12.5	370	12.5	53.5	16.5	67.8	53.1	0.839	23.6	9213	390
13		13	53.5	16.5	69.6	54.6	0.840	23.5	9470	402
14		14	53.5	16.5	73.3	57.5	0.842	23.2	9980	428
15		15	53.5	16.5	77.0	60.5	0.844	23.0	10490	455
16		16	53.5	16.5	80.7	63.5	0.846	22.8	10980	481
400 x 13	400	13	58	18	77.4	60.8	0.907	25.8	12280	476
14		14	58	18	81.4	63.9	0.908	25.5	12930	507
15		15	58	18	85.4	67.0	0.910	25.2	13580	537
16		16	58	18	89.4	70.2	0.912	25.0	14220	568
430 x 14	430	14	62.5	19.5	89.7	70.6	0.975	27.7	16460	594
15		15	62.5	19.5	94.1	73.9	0.976	27.4	17260	628
17		17	62.5	19.5	103	80.6	0.980	26.9	18860	700
20		20	62.5	19.5	115	90.8	0.986	26.3	21180	804

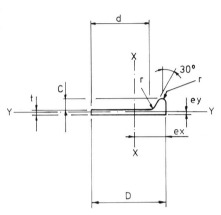

Table 4.12 Crane rails

BSC Section	Mass per metre	Rail dimension			'Cranequip' type rail clips											Lateral adjustment
		Head Width A	Base Width B	Depth D	2100 series*					8100 series†						
					C	E	F	J	K	E	F	G	H	J	K	
	Kg	mm	mm	mm	mm	mm	mm	mm	mm	mm	mm	mm	mm	mm	mm	
13 Bridge	13.31	36.0	92	47.5	–	–	–	–	–	–	–	–	–	–	–	–
16 Bridge	15.97	44.5	108	54.0	–	–	–	–	–	–	–	–	–	–	–	–
20 Bridge	19.86	50.0	127	55.5	–	–	–	–	–	–	–	–	–	–	–	–
28 Bridge	28.62	50.0	152	67.0	26	272	35	16	56	282	35	52	8	16	80	10
35 Bridge	35.38	58.0	160	76.0	32	310	40	20	70	320	40	65	10	20	100	15
50 Bridge	50.18	58.5	165	76.0	32	315	40	20	70	325	40	65	10	20	100	15
56 Crane	55.91	76.0	171	102.0	32	321	40	20	70	331	40	65	10	20	100	15
89 Crane	88.93	102.0	178	114.0	32	328	40	20	70	338	40	65	10	20	100	15
101 Crane	100.38	100.0	165	155.0	38	345	50	24	84	355	50	78	12	24	120	20
164 Crane	165.92	140.0	230	150.0	–	–	–	–	–	420	50	78	12	24	120	20

'Cranequip' crane rail clips

1. For 2100 series bolts are welded studs to girder or through bolts. Lateral resistance 100 KN per clip.

2. For 8100 series lower part of clip is welded to girder. Lateral resistance 150 KN per clip.

3. Available from CRANEQUIP LIMITED, 17a Kings Road, London SW3 4RP, U.K.

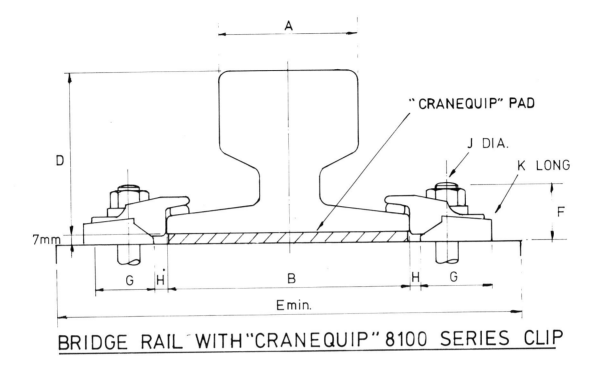

BRIDGE RAIL WITH "CRANEQUIP" 8100 SERIES CLIP

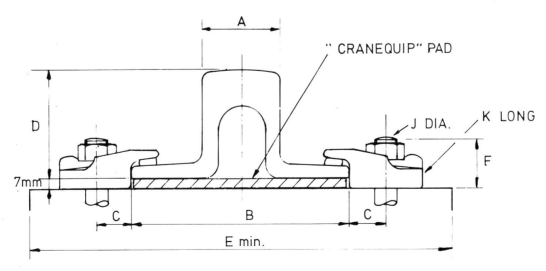

CRANE RAIL WITH "CRANEQUIP" 2100 SERIES CLIP

Table 4.13 Face clearances pitch and edge distance for bolts

Diameter		Face clearance							Min. pitch		Minimum edge distance				HSFG wrench data		
Bolt	Hole	Black bolts			HSFG bolts				Black or HSFG		BS 5950		BS 5400		Socket		Wrench
d	–	a		b	c1		c2	e		HSFG	Rolled Edge	Sheared	Black	HSFG	f	g	h
Nom	Nom	Desirable	Minimum	Min	Desirable	Minimum	Min	Min	Minimum	Desirable	*		*		Dia	Length	Overall
(12)	14	40	22	14	56	35	22	14	30	45	18	20	17	21	57	65	420
16	18	43	25	17	56	35	24	16	40	50	23	26	22	27	57	65	420
20	22	47	29	21	56	35	28	20	50	55	28	31	27	33	57	65	420
(22)	24	48	30	22	56	39	29	21	55	60	30	34	29	36	57	65	420
24	26	50	32	24	56	45	31	23	60	65	33	37	32	39	78	90	411
(27)	30	53	35	27	64	45	39	25	68	70	38	42	36	45	78	115	550
30	33	56	38	30	64	55	36	28	75	80	42	47	40	50	97	115	550
(33)	36	58	40	32	64	55	39	31	83	85	45	51	44	54	97	115	550
36	39	61	43	35	64	55	42	34	90	90	49	55	47	59	97	115	550

Notes

1. See figure illustrating face clearances.

2. Dimensions of HSFG power wrench from 'Structural Fasteners and their Application'.

3. c1 = clearance required for HSFG wrench across.
   c2 = minimum clearance with clipped washer assuming wrench other end.

4. Minimum desirable pitch for HSFG bolts based on 2 mm clearance between socket and bolt head.

5. BS 5950 edge distances demand as:
   Rolled edge – rolled, machined flame cut, sawn or planed edge.
   Sheared   – sheared or hand plane cut edge and any edge.

* These values to be used with caution because reduction in bearing capacity occurs. Distance to value for HSFG bolts. t = minimum thickness of ply.

Face clearances, pitch and edge distances for bolts figure

Maximum pitch and edge distances for bolts		BS 5400	BS 5950	
			Corrosive	Non-corrosive
Pitch	Edge distance	40 + 4t	40 + 4t	11t (M.S.) 9.7t (H.Y.S.)
	Along edge	100 + 4t or 200	16t or 200	–
	Any direction	32t or 300		–
	In direction of stress. Tension compression	16t or 200 12t or 200		14t

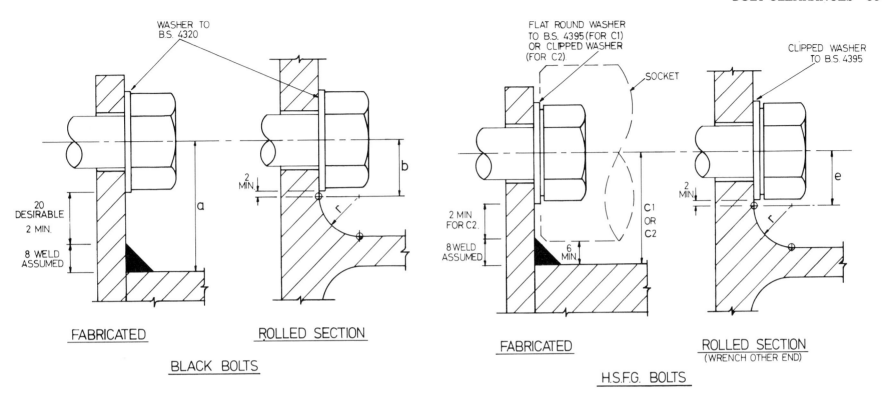

WASHER TO
B.S. 4320

20
DESIRABLE
2 MIN.

8 WELD
ASSUMED

a

2
MIN.

b

r

FABRICATED

ROLLED SECTION

BLACK BOLTS

FLAT ROUND WASHER
TO B.S. 4395 (FOR C1)
OR CLIPPED WASHER
(FOR C2)

SOCKET

CLIPPED WASHER
TO B.S. 4395

2 MIN
FOR C2.

8 WELD
ASSUMED

6
MIN.

C1
OR
C2

e

2
MIN.

r

FABRICATED

ROLLED SECTION
(WRENCH OTHER END)

H.S.F.G. BOLTS

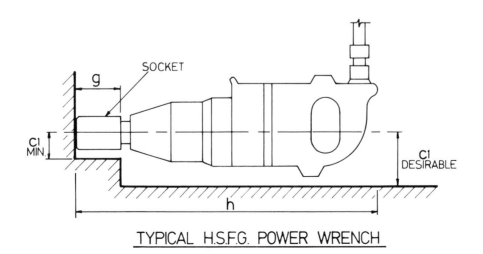

g

SOCKET

C1
MIN.

C1
DESIRABLE

h

TYPICAL H.S.F.G. POWER WRENCH

Table 4.14 Durbar floor plate

Commercial quality — tensile range between 355 and 525 N/mm² with minimum range of 77 N/mm²

BS 4360 : 43A
Lloyds and other Shipbuilding Societies Specifications. Other specifications by arrangement.

Safe uniformly distributed load in kN/m² on plates simply supported on two sides.
Extreme fibre stress : 165 N/mm².

Thickness on Plain mm.	Span in metres							
	0.6	0.8	1.0	1.2	1.4	1.6	1.8	2.0
4.5	12.29	6.97	4.47	3.10	2.28	1.77	1.37	1.12
6.0	22.06	12.41	7.94	5.52	4.04	3.12	2.44	1.98
8.0	39.24	22.12	14.09	9.83	7.18	5.54	4.34	3.56
10.0	61.22	34.45	22.00	15.33	11.22	8.67	6.78	5.55
12.5	95.82	53.91	34.44	23.99	17.56	13.57	10.61	8.70

The above safe loads include the weight of the plate.
To avoid excessive deflection, stiffeners should be used for spans over 1.1 metres.

Safe uniformly distributed loads in kg/m² on plates simply supported on four sides.
Extreme fibre stress : 16.537 kg/m².

Thickness on Plain mm.	Span in metres								Breadth Metres
	0.6	0.8	1.0	1.2	1.4	1.6	1.8	2.0	
4.5	24.86	16.36	14.05	13.21	12.85	12.70	12.58	12.53	0.60
		13.98	9.86	9.37	7.73	7.42	7.27	7.17	0.80
			8.95	6.63	5.65	5.16	4.90	4.76	1.00
6.0	44.13	29.02	24.93	23.44	22.82	22.48	22.33	22.24	0.60
		24.82	17.49	14.86	13.72	13.19	12.89	12.73	0.80
			15.89	11.77	10.02	9.15	8.70	8.44	1.00
				11.04	8.50	7.26	6.61	6.23	1.20
8.0	78.46	51.60	44.32	41.68	40.57	39.98	39.69	39.55	0.60
		44.13	31.10	26.41	24.40	23.44	22.93	22.62	0.80
			28.24	20.91	17.80	16.27	15.46	15.00	1.00
				19.60	15.11	12.89	11.74	11.08	1.20
					14.39	11.42	9.84	8.93	1.40
10.0	122.64	80.63	69.24	65.11	63.37	62.45	62.01	61.77	0.60
		68.95	48.59	41.28	38.13	36.62	35.83	35.36	0.80
			43.11	32.69	27.82	25.41	24.15	23.45	1.00
				30.63	23.61	20.15	18.34	17.28	1.20
					22.50	17.86	15.37	13.96	1.40
12.5	191.63	126.04	108.23	101.71	99.04	97.62	96.93	96.53	0.60
		107.73	75.93	64.50	59.60	57.22	56.01	55.26	0.80
			68.97	51.08	43.48	39.71	37.74	36.63	1.00
				47.88	36.92	31.51	28.67	27.05	1.20
					35.17	27.92	24.02	21.82	1.40

The above safe loads include the weight of the plate.
The deflections on the larger spans should be checked and stiffeners used if found to be necessary.

Table 4.14 *Contd*

Standard sizes and weights

Width mm	Thickness Range on Plain mm				
1000	4.5	6.0	8.0	10.0	12.5
1250					
1500					
1750					
1830	—	6.0	8.0	10.0	12.5

Consideration will be given to requirements other than standard sizes where they represent a reasonable tonnage per size, i.e. in one length and one width. Lengths up to 10,000 mm. can be supplied for plate 6 mm. thick and over.

Weights per Square Metre of Durbar Plates

Thickness on Plain (mm)*	kg/m²
4.5	37.97
6.0	49.74
8.0	65.44
10.0	81.14
12.5	100.77

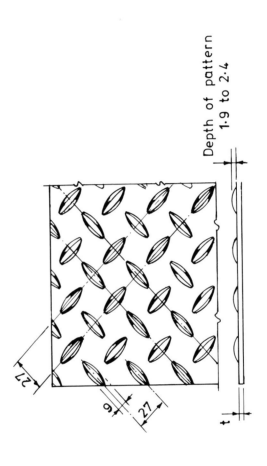

Depth of pattern 1.9 to 2.4

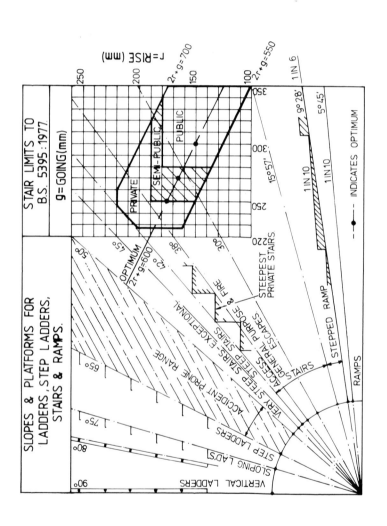

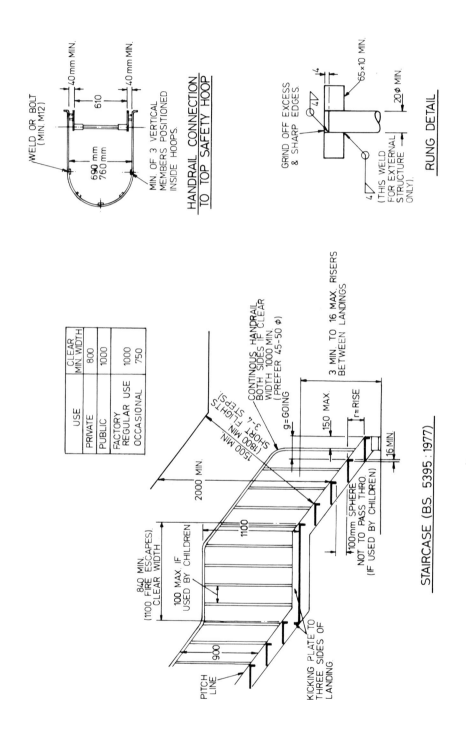

Figure 4.1 Stairs, ladders and walkways

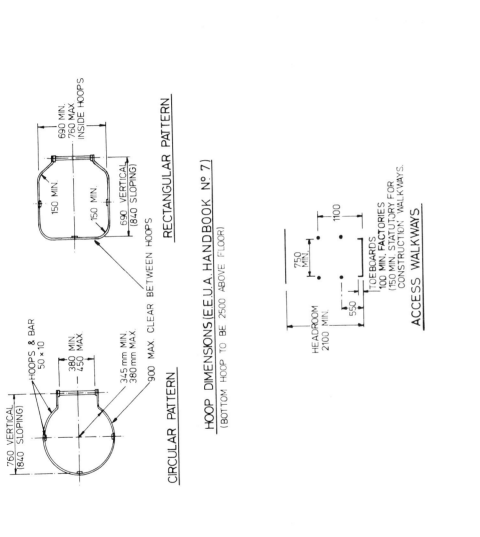

690 MIN.
760 MAX
INSIDE HOOPS

150 MIN.

150 MIN.

690 VERTICAL
(840 SLOPING)

RECTANGULAR PATTERN

900 MAX CLEAR BETWEEN HOOPS

HOOPS & BAR
50 × 10

380 MIN.
450 MAX.

345 mm MIN
380 mm MAX.

760 VERTICAL
(840 SLOPING)

CIRCULAR PATTERN

HOOP DIMENSIONS (E.E.U.A HANDBOOK Nº 7)
(BOTTOM HOOP TO BE 2500 ABOVE FLOOR)

1100

750 MIN.

550

HEADROOM
2100 MIN.

TOEBOARDS
100 MIN. FACTORIES
(150 MIN. STATUTORY FOR
CONSTRUCTION WALKWAYS.

ACCESS WALKWAYS

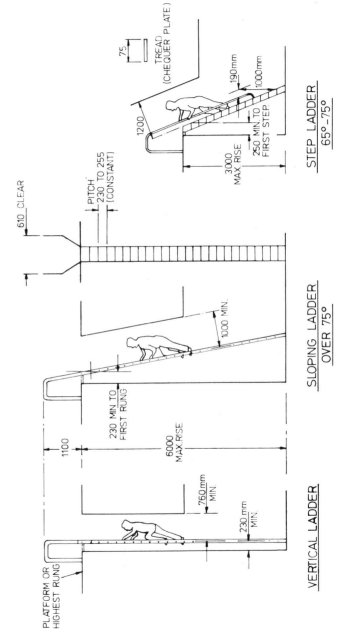

75

TREAD
(CHEQUER PLATE)

1200

190mm

1000mm

3000
MAX RISE

250 MIN TO
FIRST STEP

STEP LADDER
65° – 75°

610 CLEAR

PITCH
230 TO 255
(CONSTANT)

1000 MIN.

SLOPING LADDER
OVER 75°

230 MIN TO
FIRST RUNG

1100

6000
MAX RISE

760 mm
MIN.

230 mm
MIN.

PLATFORM OR
HIGHEST RUNG

VERTICAL LADDER

STAIRS, LADDERS AND STEP LADDERS (E.E.U.A. HANDBOOK NO. 7)

NOTE: LADDERS TO BE PROVIDED WITH HOOPS WHERE RISE EXCEEDS 2–3 m

Figure 4.1 *Contd*

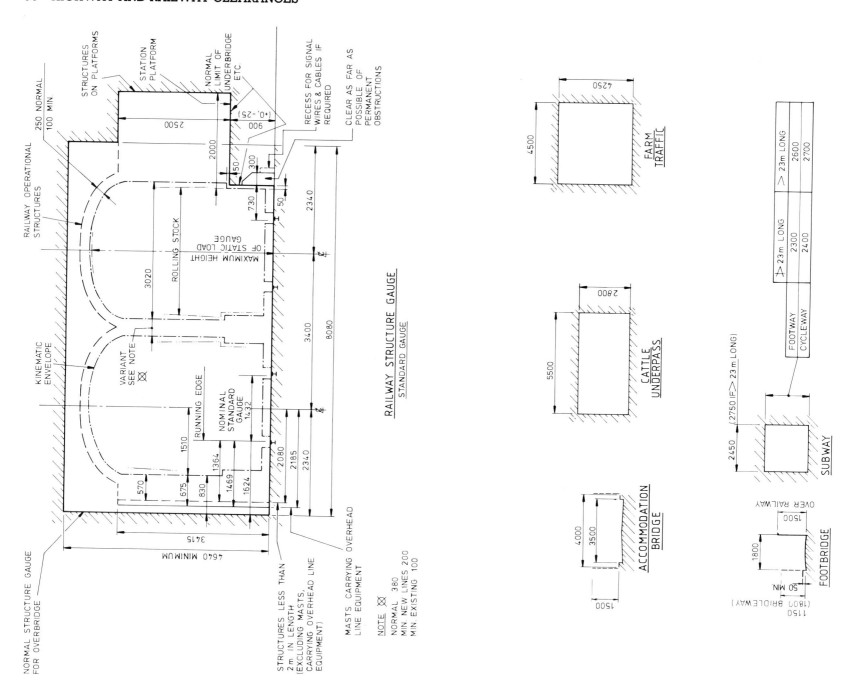

Figure 4.2 Highway and railway clearances

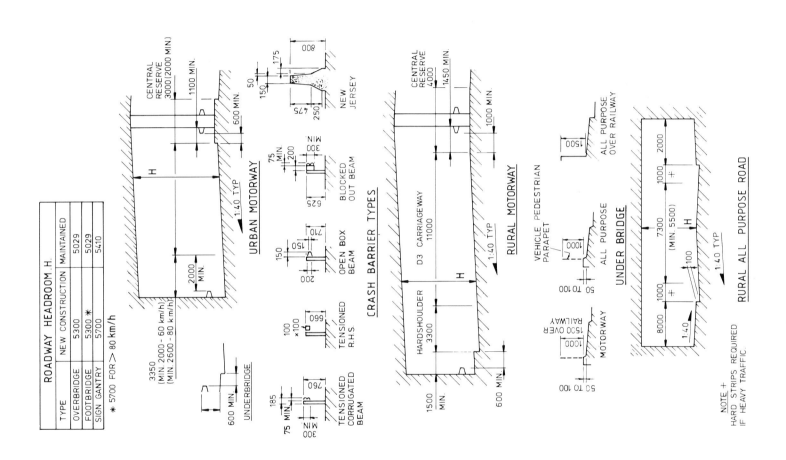

Figure 4.2 *Contd*

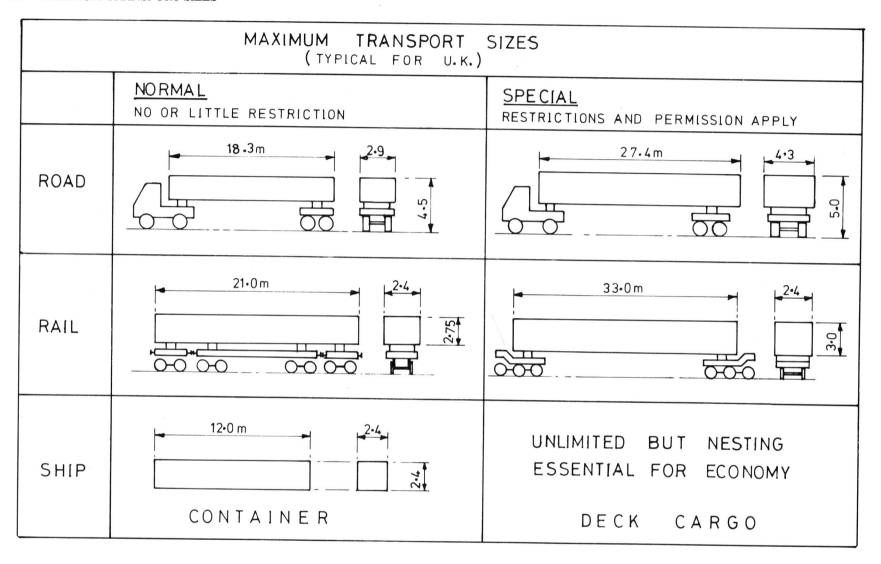

Figure 4.3 Maximum transport sizes

WELDING SYMBOLS

THESE WELDING SYMBOLS ARE BASED UPON B.S. 499 AND ARE A SELECTION OF THOSE MOST COMMONLY USED. THEY SHOULD BE USED ON ENGINEER'S & WORKSHOP DRAWINGS.

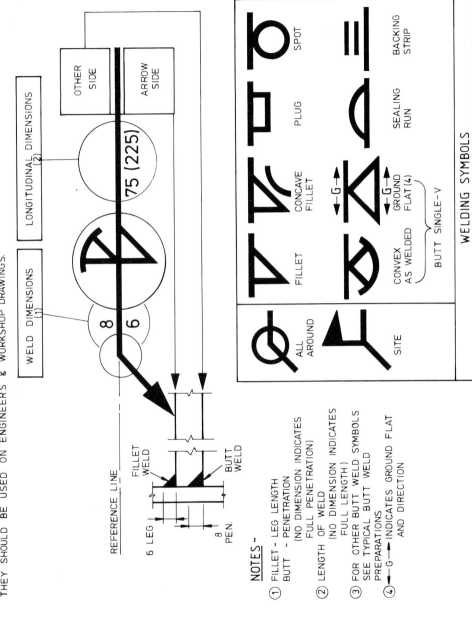

NOTES ~

① FILLET ~ LEG LENGTH
BUTT ~ PENETRATION
(NO DIMENSION INDICATES
FULL PENETRATION)

② LENGTH OF WELD
(NO DIMENSION INDICATES
FULL LENGTH)

③ FOR OTHER BUTT WELD SYMBOLS
SEE TYPICAL BUTT WELD
PREPARATIONS

④ ——G—— INDICATES GROUND FLAT
AND DIRECTION

WELDING SYMBOLS

FILLET
CONCAVE FILLET
CONVEX AS WELDED
BUTT SINGLE-V
GROUND FLAT (4)
ALL AROUND
SITE
PLUG
SPOT
SEALING RUN
BACKING STRIP

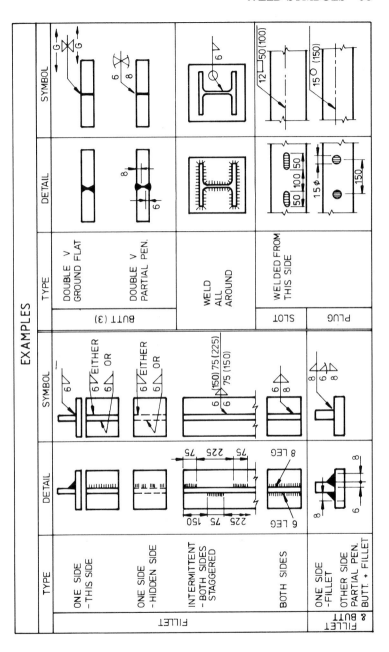

EXAMPLES

Figure 4.4 Weld symbols

TYPICAL BUTT WELD PREPARATIONS – FULL PENETRATION

THESE DETAILS ARE A TYPICAL SELECTION ONLY CONFORMING WITH THE RECOMMENDED PREPARATIONS IN BS 5135. WELD PREPARATIONS SHOULD NOT BE DETAILED ON ENGINEERS DRAWINGS BUT ARE REQUIRED ON WORKSHOP DRAWINGS

WELD & SYMBOL	DETAIL	THICKNESS T (mm)	GAP G (mm)	ANGLE α	ROOT FACE R (mm)
OPEN SQUARE BUTT 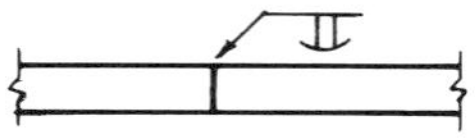	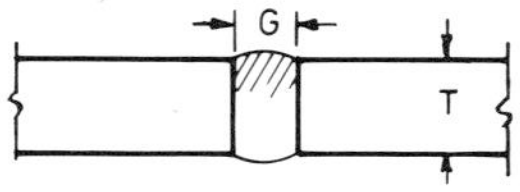	0–3 3–6	0–3 3	— —	— —
OPEN SQUARE BUTT BACKED 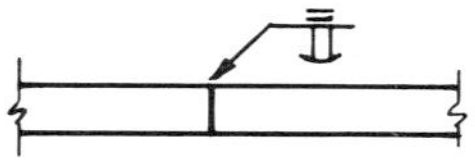	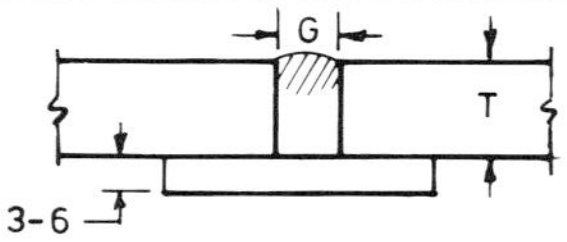	3–5 5–8 8–16	6 8 10	— — —	— — —
SINGLE V BUTT 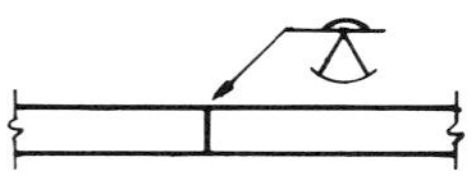	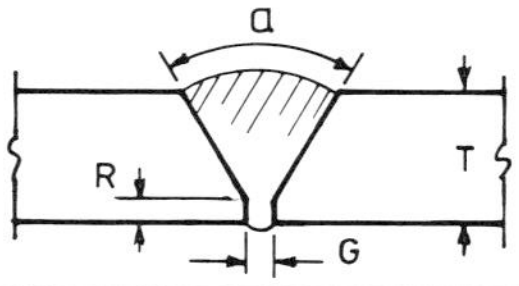	5–12 >12	2 2	60° 60°	1 2
SINGLE V BUTT BACKED 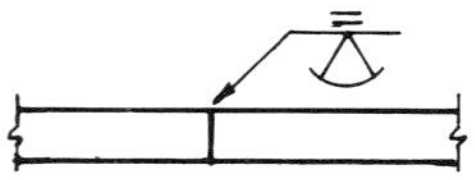	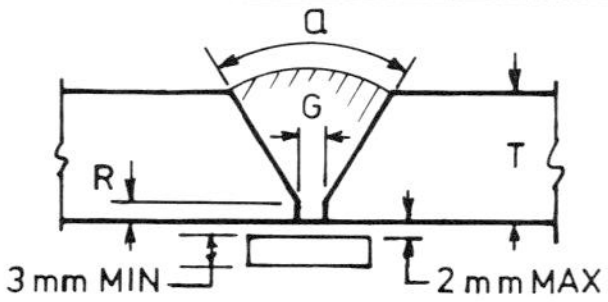	>10	6 10	45° 20°	0 0
DOUBLE V BUTT 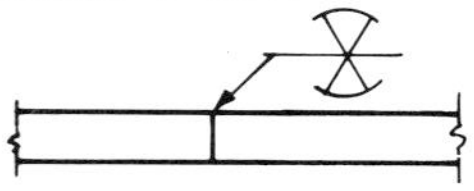	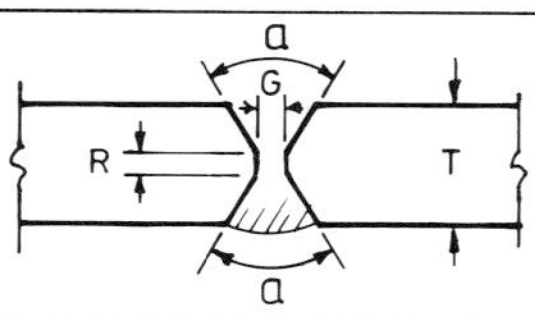	>12	3	60°	2
ASYMMETRIC DOUBLE V BUTT 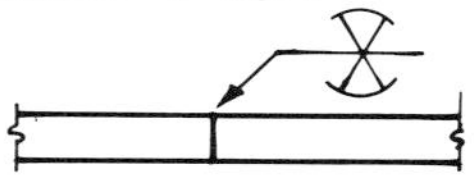	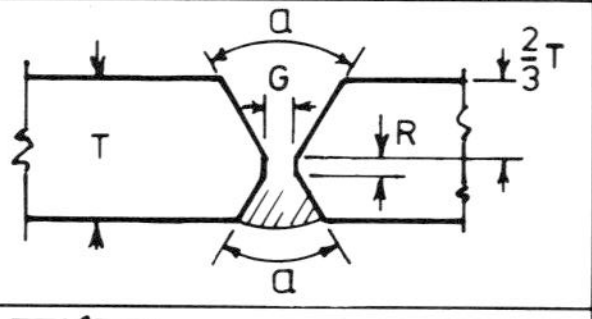	>12	3	60°	2
SINGLE J BUTT 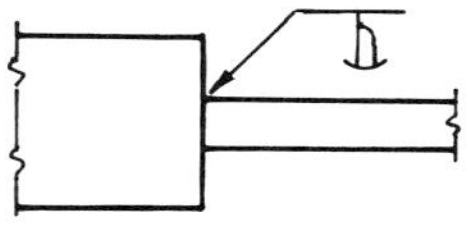	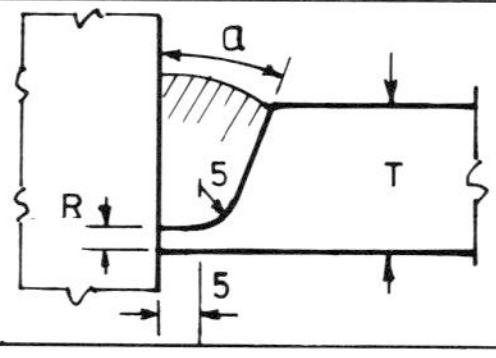	>20	—	20°	5
SINGLE U BUTT 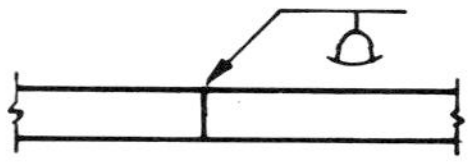	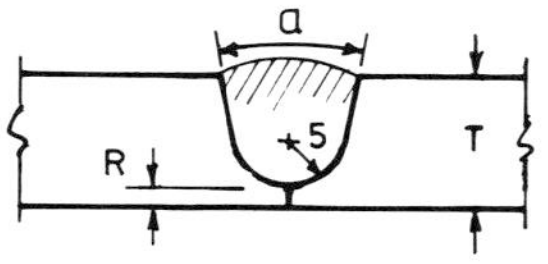	>20	—	20°	5
SINGLE BEVEL BUTT 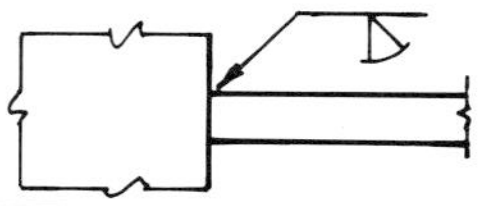	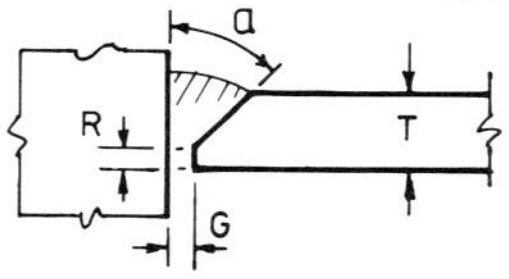	5–12 >12	3 3	45° 45	1 2
DOUBLE BEVEL BUTT 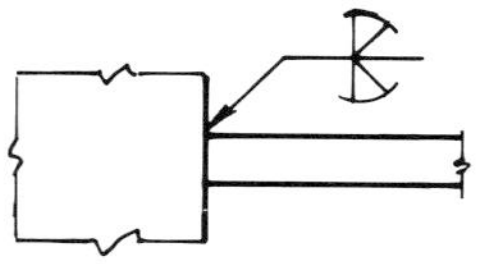	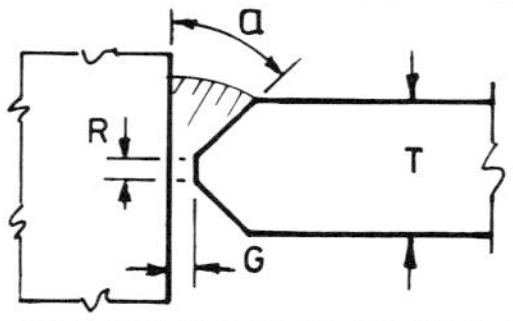	>12	3	45	2

Figure 4.5 Typical weld preparations

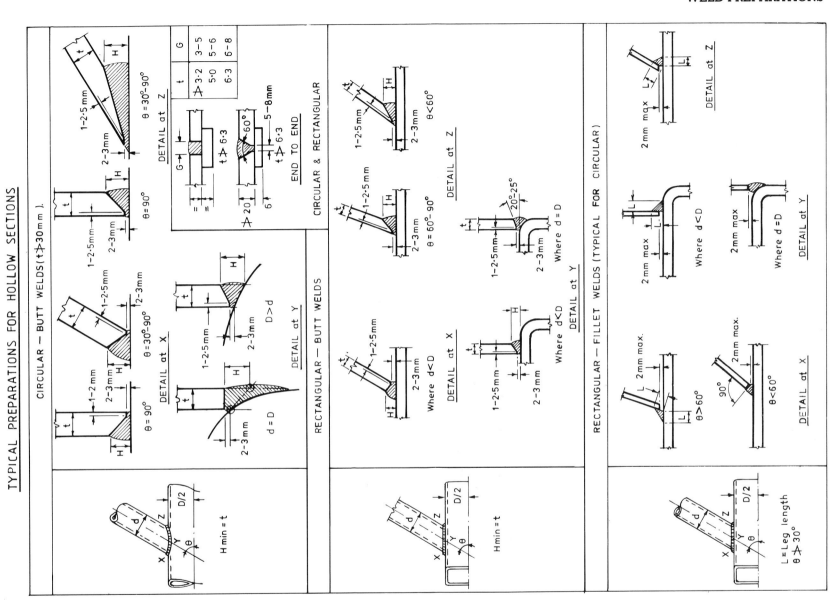

Figure 4.5 *Contd*

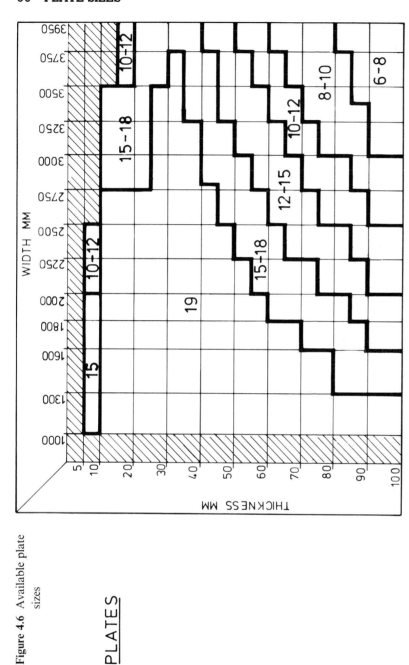

Figure 4.6 Available plate sizes

PLATES

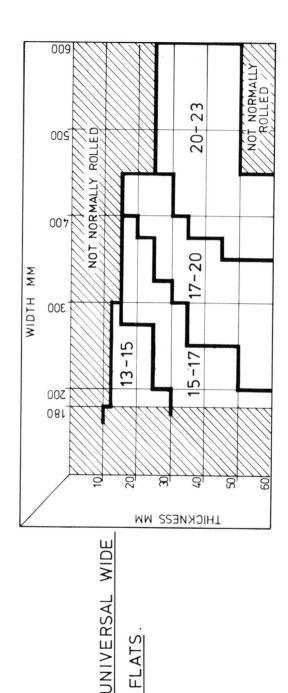

UNIVERSAL WIDE FLATS.

NOTE

THE TABLES ARE A GUIDE ONLY TO MATERIAL AVAILABLE FROM B.S.C. OTHER SIZES AND LONGER LENGTHS ARE OFTEN OBTAINABLE IN PRACTICE. IN PARTICULAR SOME UNIVERSAL WIDE FLATS MAY BE OBTAINED UPTO ABOUT 29M LONG. IT IS SUGGESTED THAT DESIGN BE BASED ON THE TABLES. BEFORE ORDERING DETAILS SHOULD BE CHECKED WITH B.S.C. PLATES.

5 TYPICAL CONNECTION DETAILS

Following are sketch examples of typical connection details. These show the principles of some of the types of connection commonly used. Both simple and rigid connections are shown as applicable to beam/column structures. A typical workshop drawing of a roof lattice girder is included in figures 5.8 and 5.9. Sketches of typical steel/timber and steel/precast concrete connections are shown in figures 5.10 and 5.11 respectively.

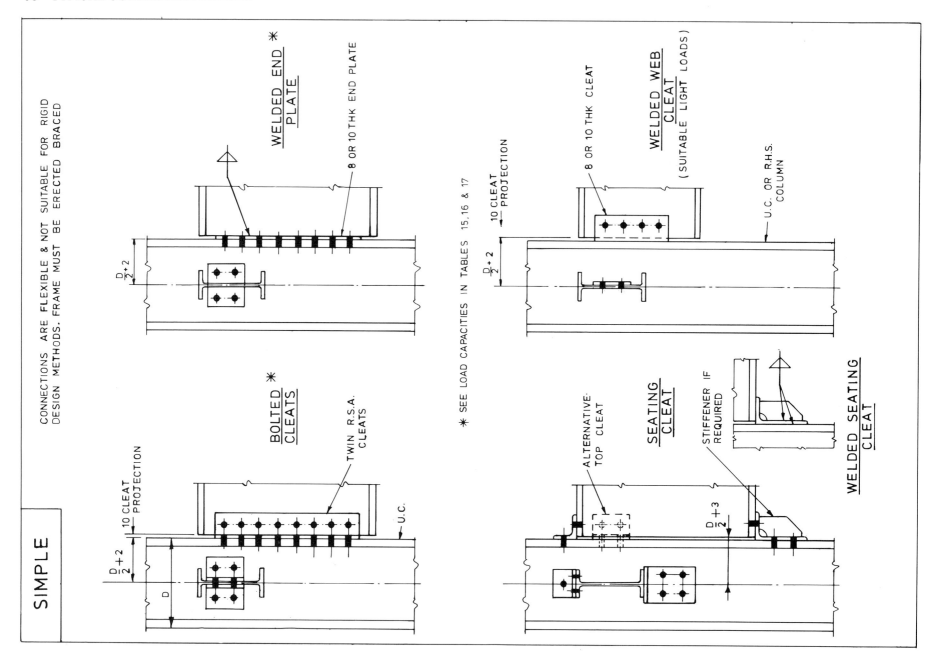

Figure 5.1 Typical beam/column connections

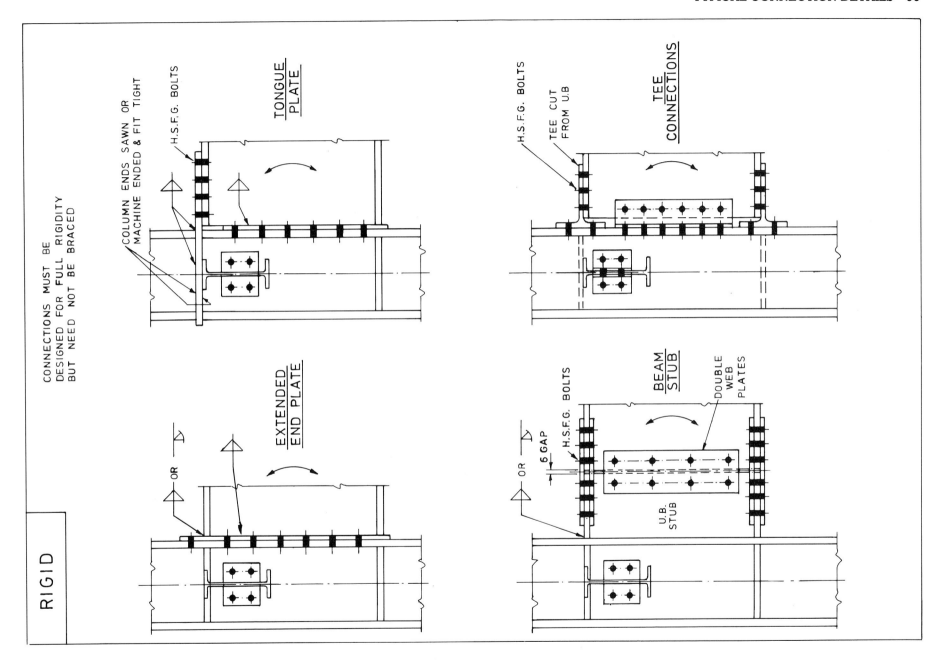

Figure 5.1 *Contd*

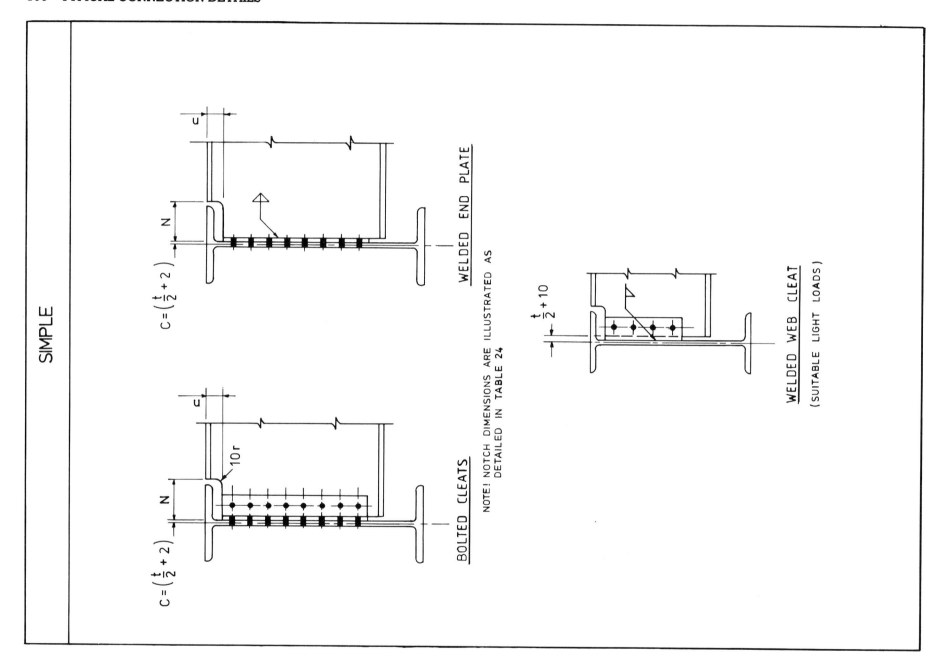

Figure 5.2 Typical beam/beam connections

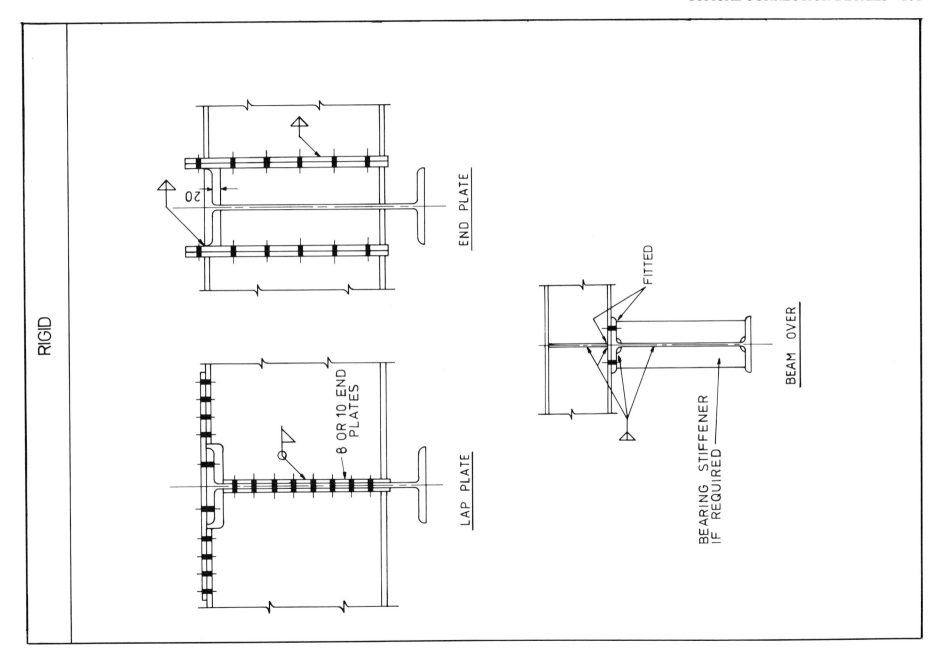

Figure 5.2 *Contd*

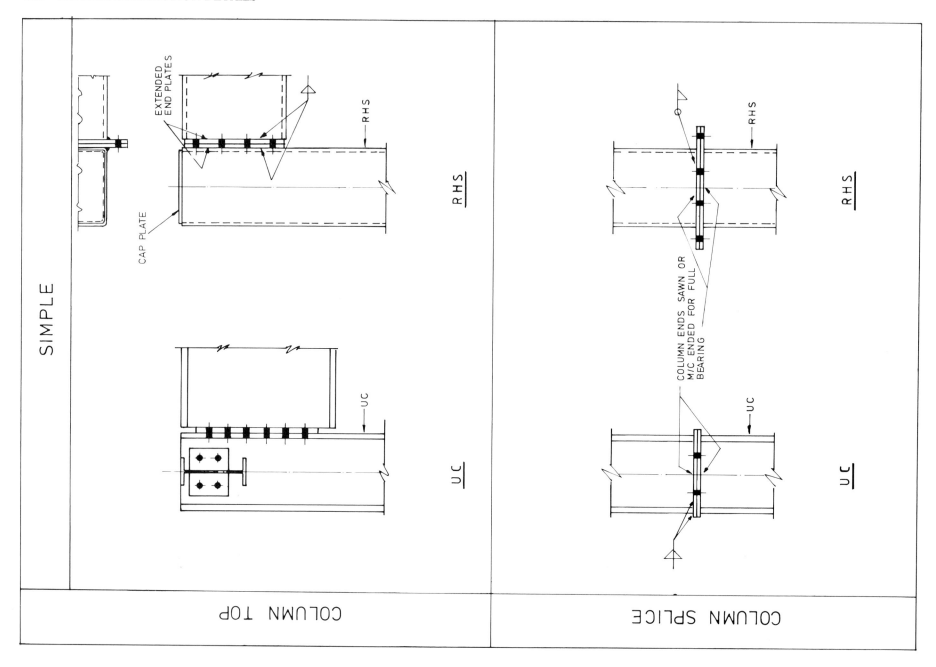

Figure 5.3 Typical column top and splice detail

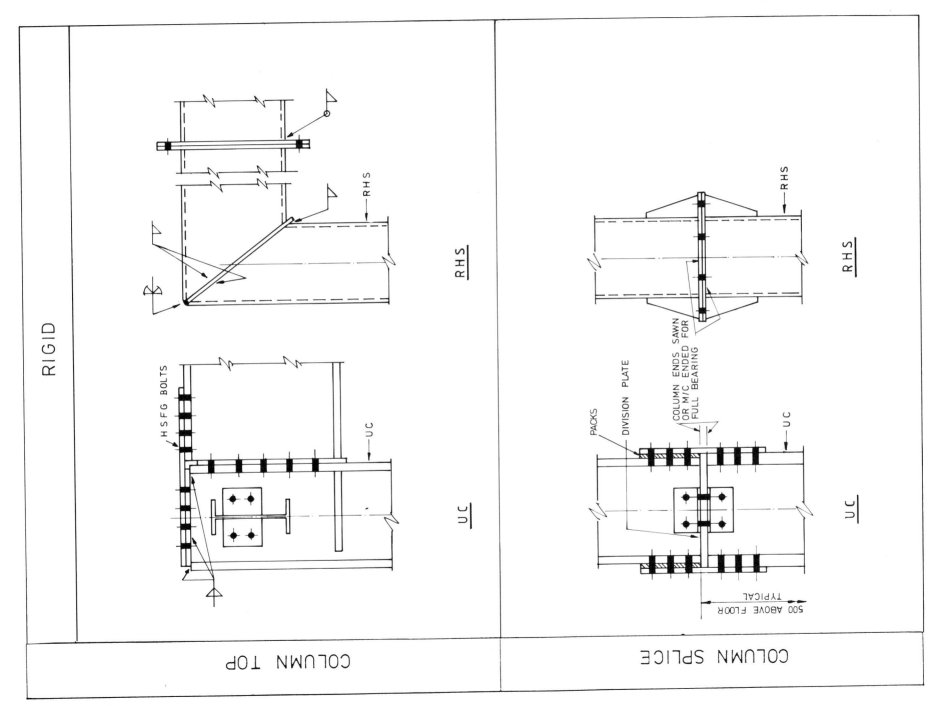

Figure 5.3 *Contd*

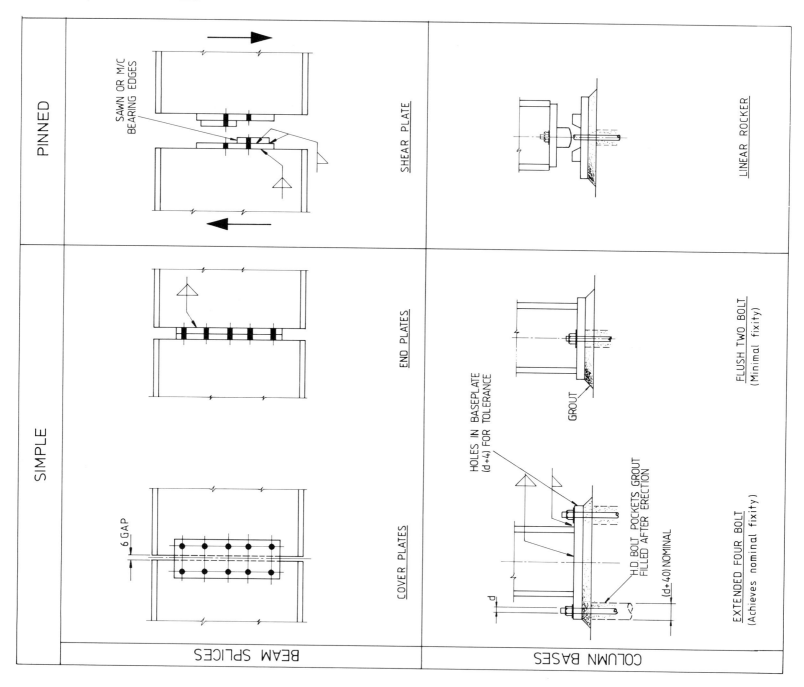

Figure 5.4 Typical beam splices and column bases

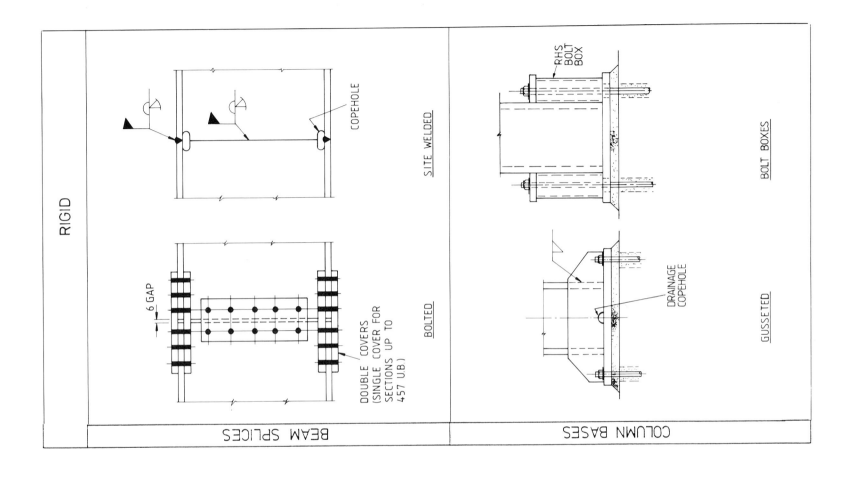

Figure 5.4 *Contd*

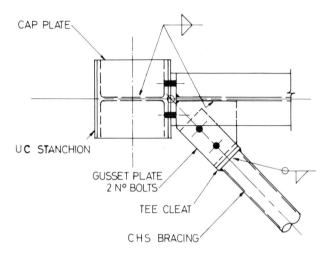

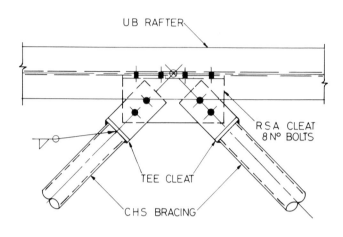

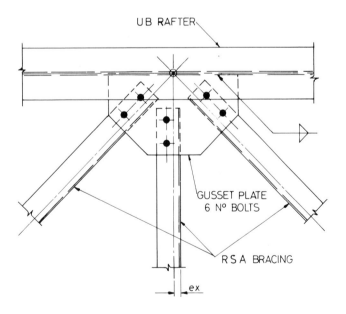

Figure 5.5 Typical bracing details

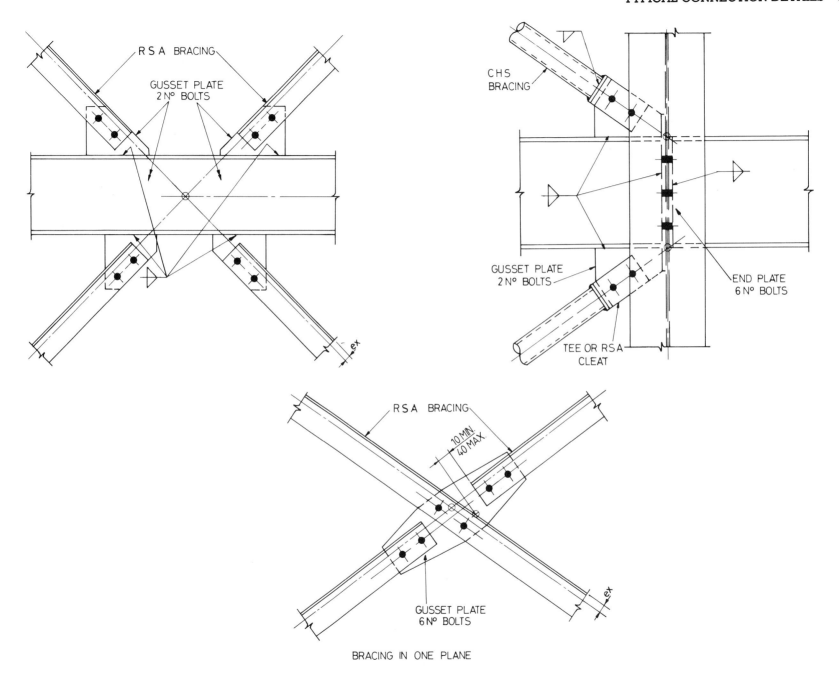

Figure 5.5 *Contd*

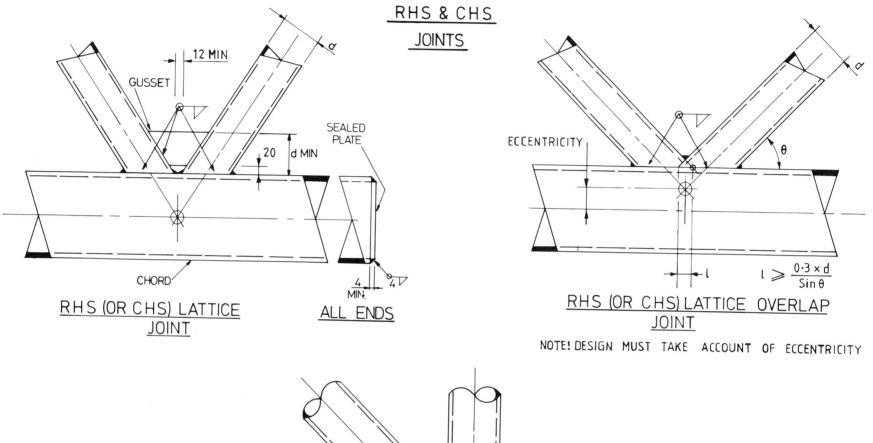

RHS & CHS JOINTS

GUSSET

12 MIN

SEALED PLATE

20   d MIN

CHORD

RHS (OR CHS) LATTICE JOINT

ALL ENDS

4 MIN   4

ECCENTRICITY

θ

$$l \geq \frac{0.3 \times d}{\sin \theta}$$

RHS (OR CHS) LATTICE OVERLAP JOINT

NOTE! DESIGN MUST TAKE ACCOUNT OF ECCENTRICITY

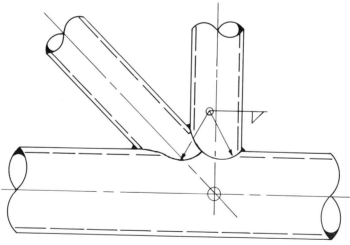

CHS (OR RHS) LATTICE OVERLAP JOINT

Figure 5.6 Typical hollow section connections

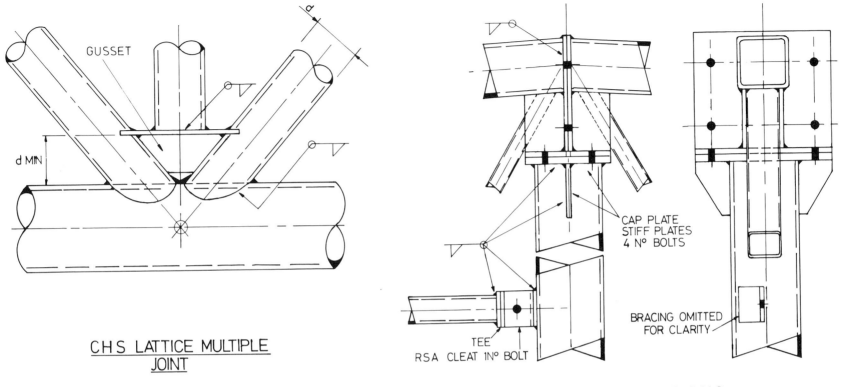

CHS LATTICE MULTIPLE
JOINT

LATTICE GIRDER TO RHS
COLUMN CAP

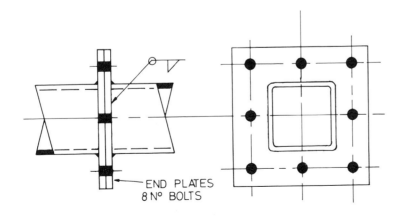

END PLATES
8 N° BOLTS

FLANGED SPLICE
JOINT

NOTE: ALL HOLLOW SECTIONS TO BE
FULLY SEALED BY WELDING.

Figure 5.6 *Contd*

WELDED

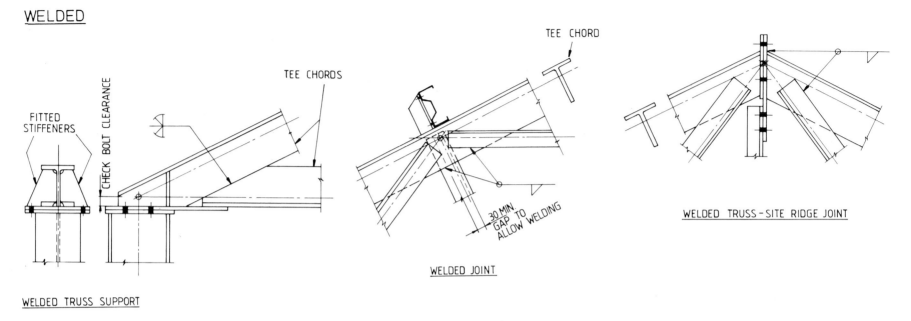

FITTED STIFFENERS

CHECK BOLT CLEARANCE

TEE CHORDS

WELDED TRUSS SUPPORT

TEE CHORD

30 MIN. GAP TO ALLOW WELDING

WELDED JOINT

TEE CHORD

WELDED TRUSS–SITE RIDGE JOINT

TRUSS JOINTS

BOLTED

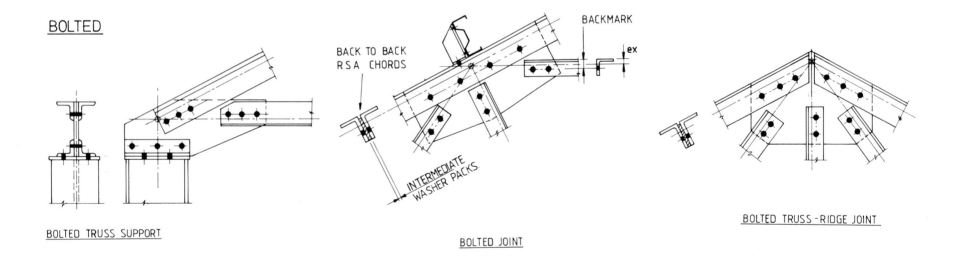

BOLTED TRUSS SUPPORT

BACK TO BACK RSA CHORDS

INTERMEDIATE WASHER PACKS.

BOLTED JOINT

BACKMARK

ex

BOLTED TRUSS–RIDGE JOINT

Figure 5.7 Typical truss details

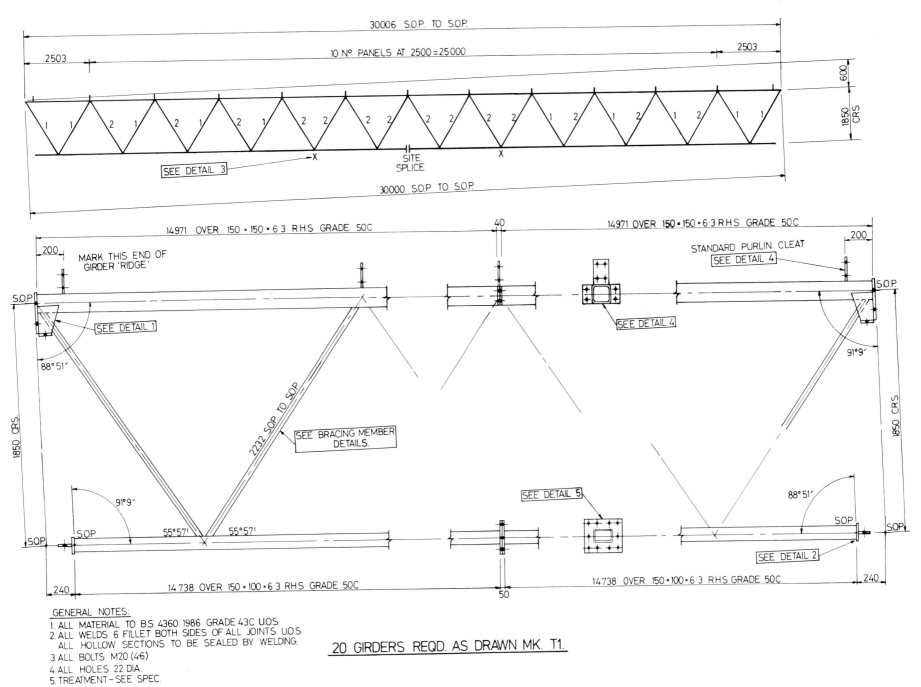

Figure 5.8 Workshop drawing of lattice girder – 1

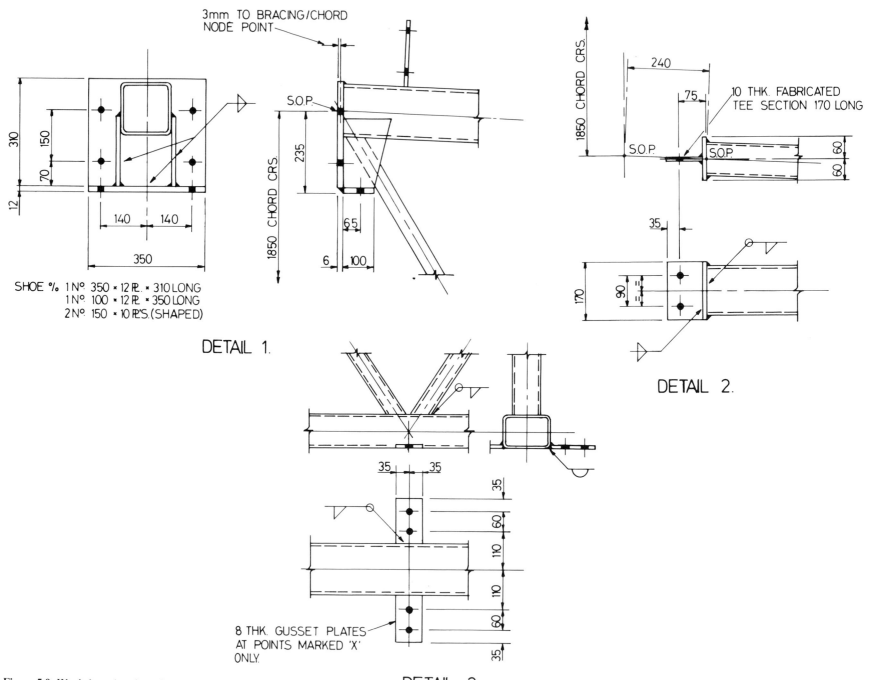

SHOE % 1 Nº. 350 × 12 PL. × 310 LONG
1 Nº. 100 × 12 PL. × 350 LONG
2 Nº. 150 × 10 PL'S.(SHAPED)

DETAIL 1.

DETAIL 2.

8 THK. GUSSET PLATES
AT POINTS MARKED 'X'
ONLY.

DETAIL 3.

Figure 5.9 Workshop drawing of lattice girder – 2

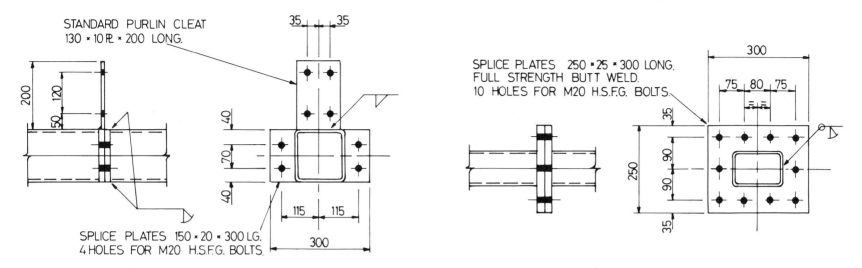

STANDARD PURLIN CLEAT
130 × 10 Pℓ × 200 LONG.

SPLICE PLATES 150 × 20 × 300 LG.
4 HOLES FOR M20 H.S.F.G. BOLTS.

DETAIL 4.

SPLICE PLATES 250 × 25 × 300 LONG.
FULL STRENGTH BUTT WELD.
10 HOLES FOR M20 H.S.F.G. BOLTS.

DETAIL 5.

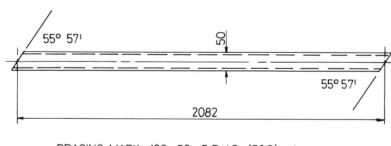

BRACING MARK 100 × 50 × 5 R.H.S. (50C) – 1.
BRACING MARK   90 × 50 × 5 R.H.S. (50C) – 2.

NOTE !
    FOR FABRICATION WORKS WITH TEMPLATING
    FACILITY THIS DETAIL NOT NECESSARY.

Figure 5.9 *Contd*

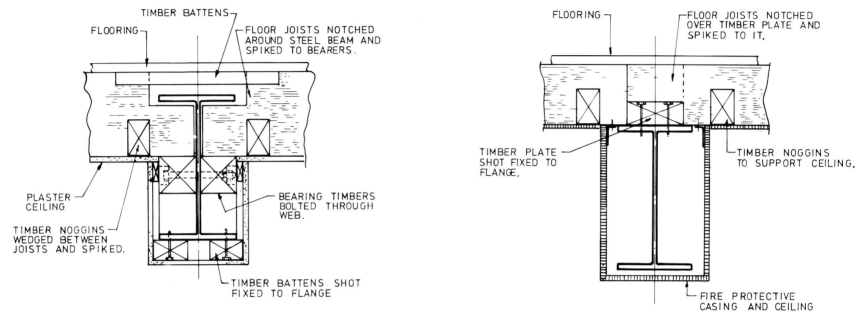

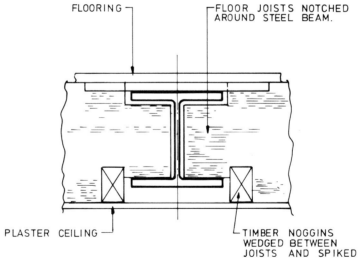

U.C. SECTION USED AS BEAM
TO PROVIDE FLUSH SOFFIT.

Figure 5.10 Typical steel/timber connections

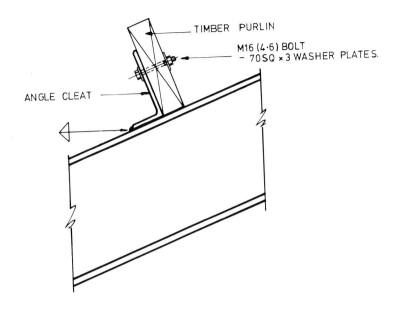

TIMBER PURLIN

M16 (4·6) BOLT
- 70 SQ × 3 WASHER PLATES.

ANGLE CLEAT

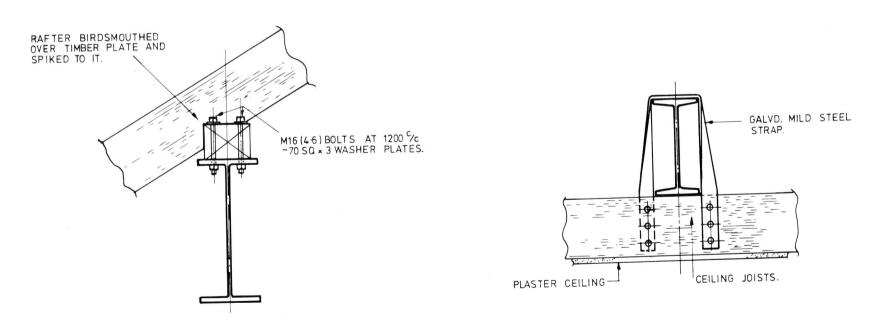

RAFTER BIRDSMOUTHED
OVER TIMBER PLATE AND
SPIKED TO IT.

M16 (4·6) BOLTS AT 1200 $^c/_c$
-70 SQ × 3 WASHER PLATES.

GALVD. MILD STEEL
STRAP.

PLASTER CEILING

CEILING JOISTS.

Figure 5.10 *Contd*

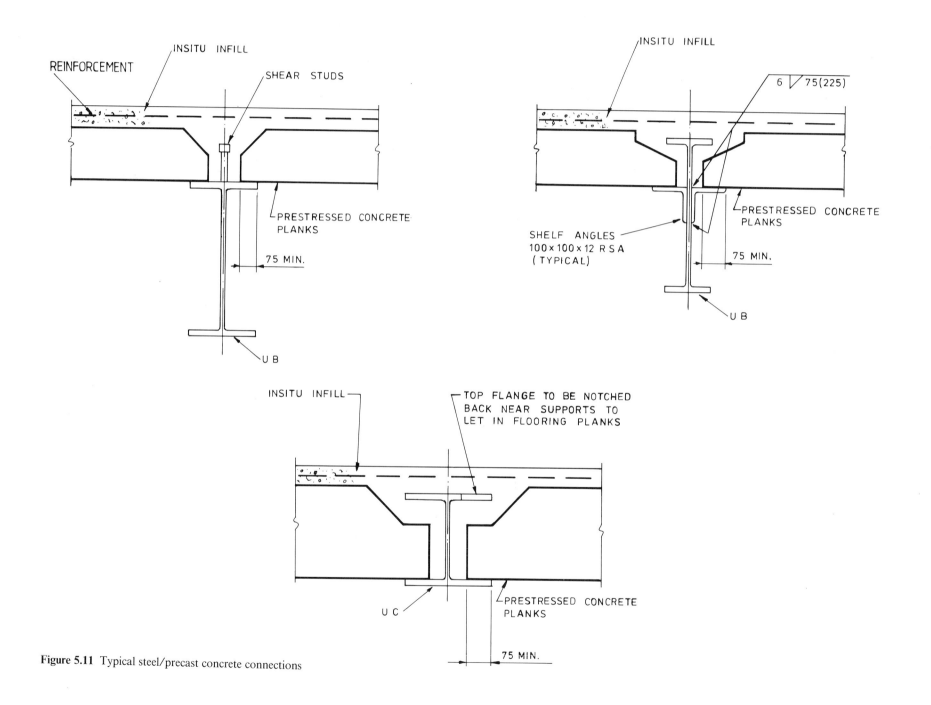

Figure 5.11 Typical steel/precast concrete connections

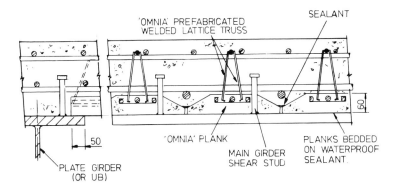

PART PRECAST DECK USING
'OMNIA' SYSTEM

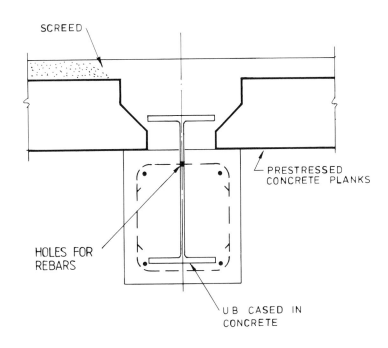

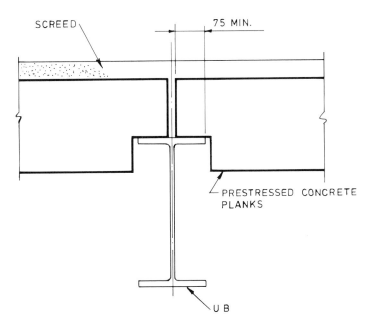

Figure 5.11 *Contd*

6 EXAMPLES OF STRUCTURES

Following are examples of various types of structures utilising structural steelwork. Some of these are taken from actually constructed projects designed by the authors. The practices and details shown will be suitable for many countries of the world. The member sizes are as actually used where shown, but it is emphasised that they might not always be appropriate in a particular case, because of variations in loading or requirements of different design codes.

A brief description of each structure type is included giving particular reasons for use and any particular influences which affect the method of construction or details employed.

6.1 MULTI-STOREY FRAME BUILDINGS

Multi-storey steel frames provide the structural skeleton from which many commercial and office buildings are supported. Steel has the advantage of being speedy to erect and it is very suitable in urban situations where conditions are restrictive. This is further exploited by the use of rapidly constructed floors and claddings. This means that a 'dry envelope' is available at the earliest possible date so that interior finishes can be advanced and the building occupied sooner. Floor systems used include precast concrete and composite profiled galvanised metal decking which can also be made composite with the steel frame. Such decking is supplied in lengths which span over several secondary beams and shear studs are then welded through it. Mesh reinforcement is provided to prevent cracking of the concrete slab.

The structural layout of beams and columns will largely depend upon the required use. Modern buildings require extensive services to be accommodated within floors and this may dictate that beams contain openings. Here castellated or tapered beams can be useful. In general, floors are supported by secondary and main beams usually of universal beams, supported by columns formed from UCs. The spacing of secondary beams is dictated by the floor type, typically 2.5 m to 3.5 m. An important design decision is whether stability against horizontal forces (e.g. due to wind or earthquake) is to be resisted using rigid connections or whether bracing is to be supplied and simple connections used. Alternatively other elements may be available such as lift shafts or shear walls, allied with the lateral rigidity of floors, to which the steelwork can be secured. In this case temporary stability may need to be supplied using diagonal bracings during erection until a means of permanent stability is provided.

The example shown in figures 6.1.1 to 6.1.5 is a two-storey office building with floors and roof of composite profiled steel decking. Beam to column connections are of simple type and stability is provided by wind bracings installed within certain external walls. Because there are only two storeys the columns are fabricated full height without splices. The top of the columns can be detailed to suit future upward extension if required. Connections for the cantilevered canopy beams are of rigid end plate type.

Figure 6.1.2 is a first floor part plan being part of the engineer's drawings which gives member sizes and ultimate limit state beam reactions for the fabricator to design the connections. Typical connections are shown in figure 6.1.3. Workshop drawings of a beam and a column are shown in figures 6.1.4 and 6.1.5 respectively which are prepared by the fabricator after designing the connections.

Fire resistance

Generally, multi-storey steel framed buildings are required by Building Regulations to exhibit a degree of fire resistance that is dependent on the building form and size. Fire resistance is specified as a period of time, e.g. $\frac{1}{2}$ hour, 1 hour, 2 hours etc. and is normally achieved by insulation in a form of cladding. The thickness of cladding required is therefore dependent on material type and period of resistance. Traditional materials such as concrete, brickwork and plasterboard are still used but have to a great extent been replaced by modern lightweight materials such as

vermiculite and mineral fibre. Asbestos is no longer used for health reasons.

Lightweight claddings are available in spray form or board; sprays, being unsightly, are generally used where they will not be seen, e.g. floor beams behind suspended ceilings. Boards can be prefinished or decorated and are fixed typically by screwing mainly to noggins or wrap around steel straps. Typical arrangements are shown in figure 6.1.6. The thickness of cladding and fixing clearly affects building details and therefore warrants early consideration [24].

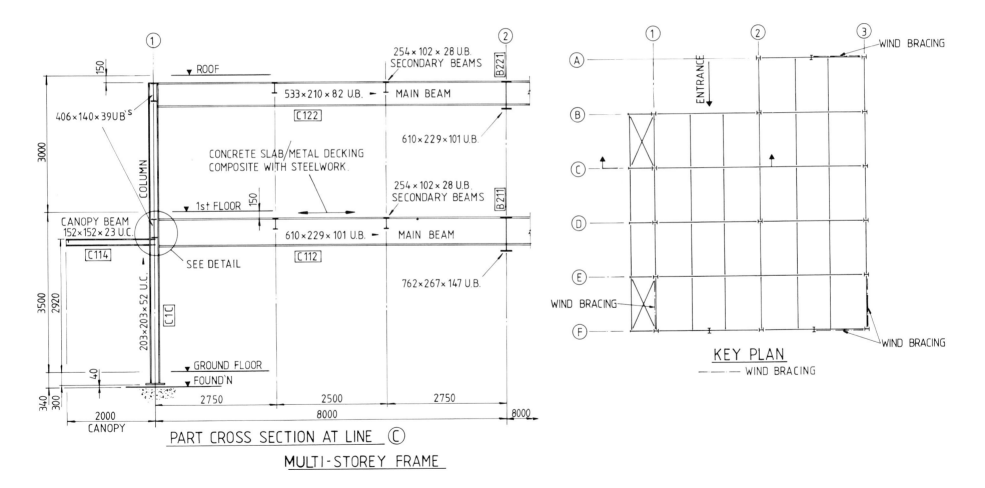

PART CROSS SECTION AT LINE Ⓒ

MULTI-STOREY FRAME

KEY PLAN
— · — WIND BRACING

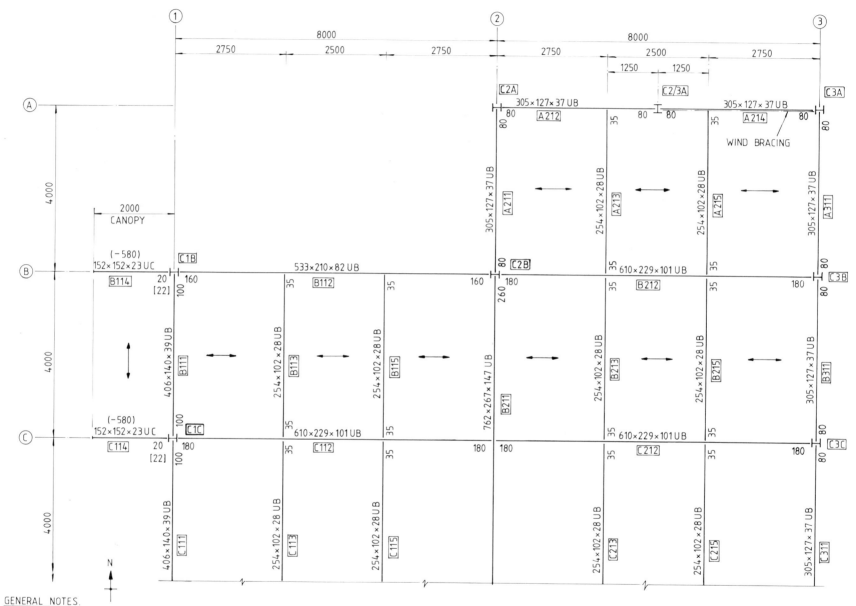

GENERAL NOTES:
1. REACTIONS ARE FACTORED LOADS TO B.S.5950. IN K.N. BENDING MOMENTS IF ANY IN K.N.M IN BRACKETS E.G. [180]
2. ALL STEEL TO B.S.4360:1986 GRADE 43A.
3. ALL BEAM MARKS TO BE AT NORTH OR EAST END. ALL COLUMN MARKS TO BE ON FLANGE FACING NORTH OR EAST.
4. ALL BEAMS AT 150 BELOW 1st FLOOR EXCEPT WHERE SHOWN IN BRACKETS E.G. (−580)
5. ◄──► INDICATES THE DIRECTION OF METAL DECKING OR CLADDING.

STEELWORK
FIRST FLOOR PLAN

Figure 6.1.2 Multi-storey frame building

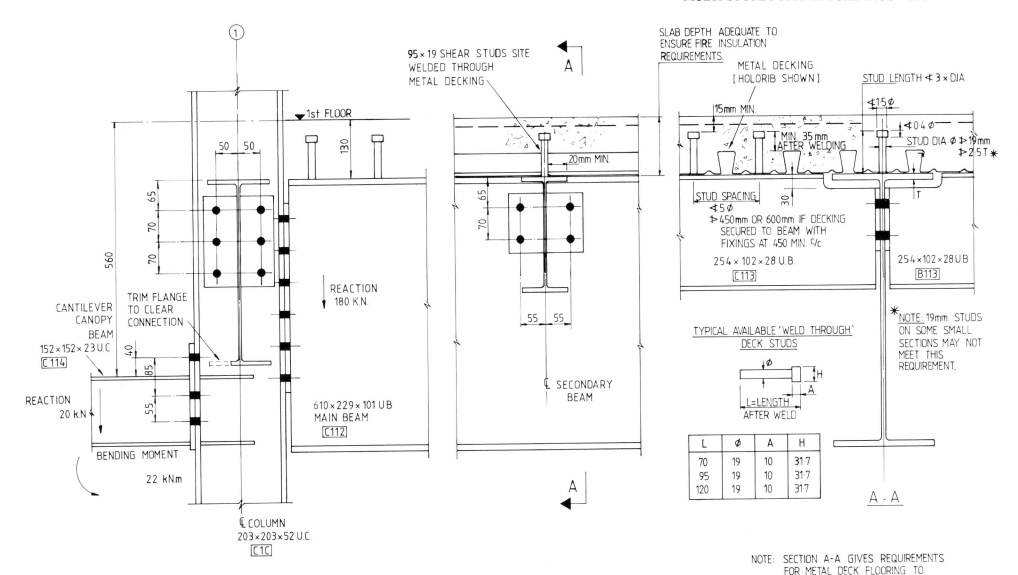

95 × 19 SHEAR STUDS SITE
WELDED THROUGH
METAL DECKING

SLAB DEPTH ADEQUATE TO
ENSURE FIRE INSULATION
REQUIREMENTS.

METAL DECKING
[HOLORIB SHOWN]

STUD LENGTH ≮ 3 × DIA

STUD DIA Φ ⊅ 19mm
⊅ 2·5 T *

1st FLOOR

REACTION
180 K.N.

TRIM FLANGE
TO CLEAR
CONNECTION

CANTILEVER
CANOPY
BEAM
152 × 152 × 23 U.C.
C 114

REACTION
20 k.N

BENDING MOMENT

22 kN.m

℄ COLUMN
203 × 203 × 52 U.C.
C1C

610 × 229 × 101 UB
MAIN BEAM
C112

℄ SECONDARY
BEAM

STUD SPACING
≮ 5 Φ
⊅ 450mm OR 600mm IF DECKING
SECURED TO BEAM WITH
FIXINGS AT 450 MIN. C/c

254 × 102 × 28 U.B.
C113

254 × 102 × 28 U.B.
B113

15mm MIN.

MIN. 35 mm
AFTER WELDING

≮ 1·5 Φ
≮ 0·4 Φ

* NOTE: 19mm STUDS
ON SOME SMALL
SECTIONS MAY NOT
MEET THIS
REQUIREMENT.

TYPICAL AVAILABLE 'WELD THROUGH'
DECK STUDS

L = LENGTH
AFTER WELD

L	Φ	A	H
70	19	10	31·7
95	19	10	31·7
120	19	10	31·7

A - A

NOTE: SECTION A-A GIVES REQUIREMENTS
FOR METAL DECK FLOORING TO
B.S. 5950 PART 4 AND FOR
THROUGH DECK STUD WELDING.

TYPICAL CONNECTIONS AT FIRST FLOOR
GRID REF ①-℃.

Figure 6.1.3 Multi-storey frame building

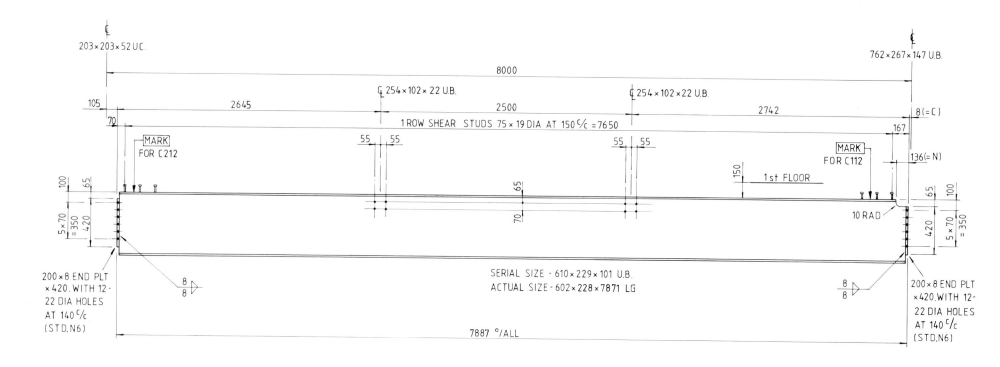

1-BEAM REQ'D AS DRAWN & NOTED MK'D C112
1-BEAM REQ'D AS DRAWN & NOTED MK'D C212

GENERAL NOTES
UNLESS OTHERWISE STATED
1. ALL MATERIAL TO BS 4360: GRADE 43A.
2. ALL BOLTS M20 GRADE 4·6
3. ALL HOLES 22 DIA
5. TREATMENT – SEE SPEC

WORKSHOP DRAWING - BEAM DETAIL

Figure 6.1.4 Multi-storey frame building

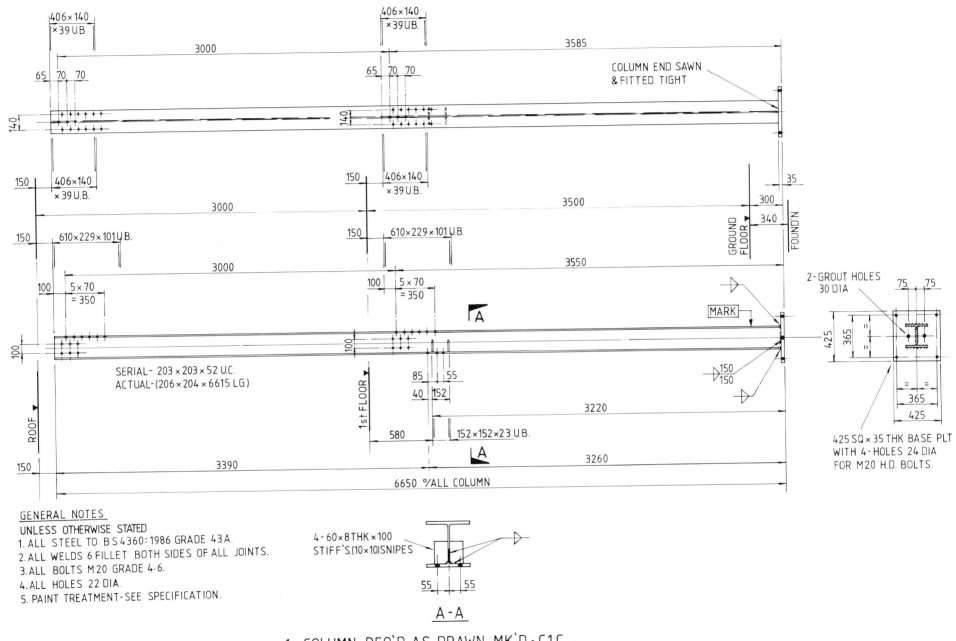

GENERAL NOTES
UNLESS OTHERWISE STATED
1. ALL STEEL TO BS4360:1986 GRADE 43.A.
2. ALL WELDS 6 FILLET BOTH SIDES OF ALL JOINTS.
3. ALL BOLTS M20 GRADE 4.6.
4. ALL HOLES 22 DIA.
5. PAINT TREATMENT-SEE SPECIFICATION.

4-60×8THK×100
STIFF'S.(10×10)SNIPES

A-A

1-COLUMN REQ'D AS DRAWN MK'D-C1C

WORKSHOP DRAWING - COLUMN DETAIL

Figure 6.1.5 Multi-storey frame building

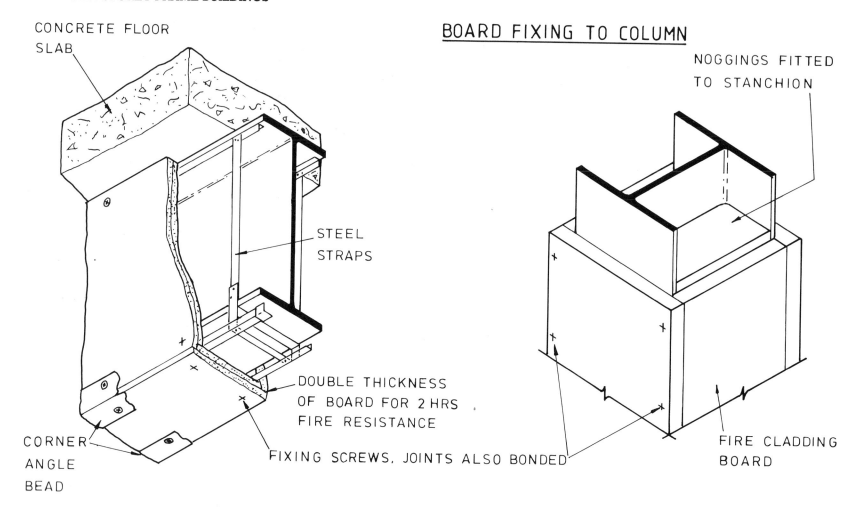

CONCRETE FLOOR SLAB

BOARD FIXING TO COLUMN

NOGGINGS FITTED TO STANCHION

STEEL STRAPS

DOUBLE THICKNESS OF BOARD FOR 2 HRS FIRE RESISTANCE

CORNER ANGLE BEAD

FIXING SCREWS, JOINTS ALSO BONDED

FIRE CLADDING BOARD

BOARD FIXING TO FLOOR BEAM

Figure 6.1.6 Multi-storey frame building

6.2 SINGLE STOREY FRAME BUILDINGS

Single storey frame buildings are extensively used for industrial, commercial and leisure buildings. In many countries of the world they are economically constructed in steel because the principal loads, namely the roof and wind are relatively light, yet the spans may be large, commonly up to about 45 m. Steel with its high strength:weight characteristics is ideally suited. The frame efficiently carries the roof cladding independently of the walls thus offering flexibility in location of openings or partitions. Side cladding is directly attached to the frame which gives stability to the whole building. This system is also ideally suited to structures in seismic areas. Sometimes solid side cladding such as brickwork is used part or full height, and it is often convenient to stabilise this by attachment to the frame although vertical support is independent. Generally the steel frame terminates at least 300 mm below floor level upon its own foundations. This permits flexibility in future use of the floor which may need to contain openings or basements and be replaced periodically if subjected to heavy use. Any internal walls or partitions are generally not structurally connected to the frame so that there is flexibility in relocation for any different future occupancy.

Figure 6.2.2 shows a number of frame types. A single bay is indicated but multiple bays are often used for large buildings for economy when internal columns are permitted. Portal frames, the most common type, are described in section 6.4.

Requirements for natural lighting by provision of translucent sheeting or glazing often govern roof shape and therefore the type of frame. In particular the monitor roof type (Figure 6.8(j)) provides a high degree of natural light. The wide use of lightweight claddings, especially profiled steel sheeting (usually galvanised and plastic coated in a range of colours) which have largely displaced other materials, permits economic roofs of shallow pitch (typically 1:10 or 6°). Such cladding is available with insulation layer, which can, if necessary, be incorporated below purlin level to produce a flush interior if needed for hygenic reasons. Flat roofs, but with provision for drainage falls, covered by proprietary roof decking are also used, but at generally greater expense. Sufficient camber or crossfall must be used to ensure rainwater run-off. Depending upon the required use, provision of a suspended ceiling may also decide the frame type. For industrial buildings internal cranes are usually required in the form of electric overhead travelling (EOT) type supported by gantry girders mounted on the frame. Clearances and wheel loads for the crane (or cranes) must be considered, which will vary according to the particular manufacturer.

The structural form most generally used is the portal frame described in section 6.3. Figure 6.2.2 shows a number of other types. The stanchions and truss type frames (a) and (c) are more suited to roofs having pitch greater than 3:10. Presence of the bottom tie is convenient for support of any suspended ceilings, but a disadvantage is that the stanchion bases must be fixed to ensure lateral stability. The lattice stanchion and truss frame (d) is suitable for EOT cranes exceeding 10 tonnes capacity. Where

appearance of the frame is important or where industrial processes demand clean conditions, hollow section members are suitable using triangular lattice girders as (g) or space grids (h). The latter are uneconomic for spans up to about 40 m, but are suitable for long spans if internal stanchions are not permitted.

Bolted site connections are generally necessary between stanchions and roof structure with the latter fabricated full span length where delivery allows. Truss or lattice roofs usually have welded workshop connections. Secondary members in the form of sheeting rails or purlins are usually of cold formed sections (see section 6.3). A vital consideration is longitudinal stability, especially during erection, which requires the provision of bracing to walls taking account of the location of side openings. Roof bracing is also necessary except where plan rigidity is inherent such as with a space grid. Gantry girders for EOT cranes should incorporate details which permit adjustment to final position as shown in figure 6.2.1, and possible replacement of rails during the life of the structure. Safety requirements such as space for personnel between end of crane to structures and positioning of power cables must be met.

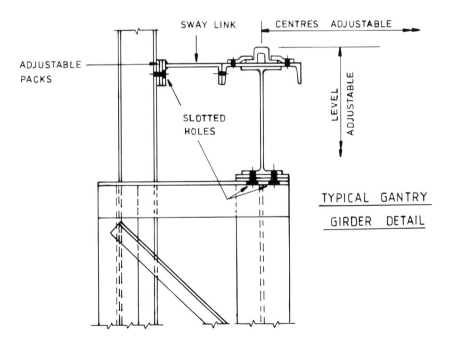

Figure 6.2.1 Single storey frame building

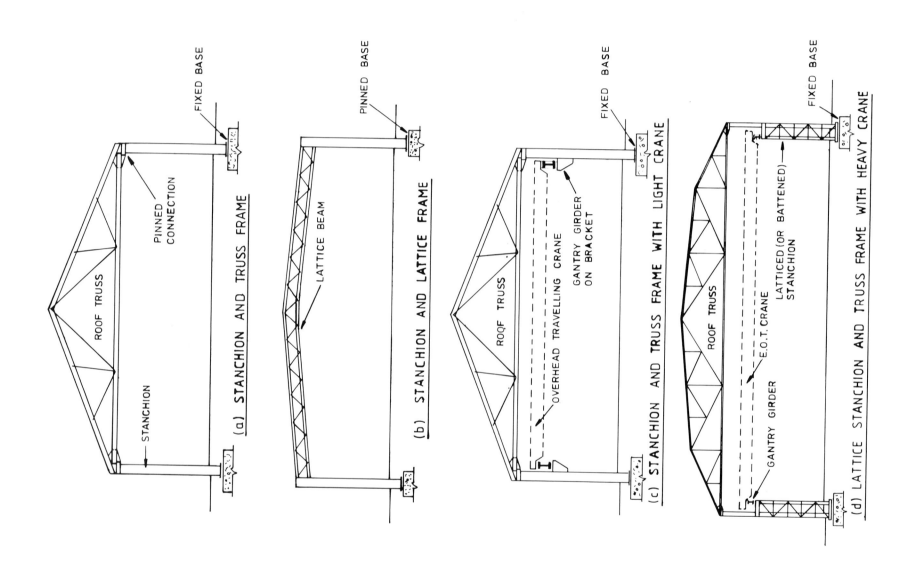

FIXED BASE

PINNED CONNECTION

ROOF TRUSS

STANCHION

(a) **STANCHION AND TRUSS FRAME**

PINNED BASE

LATTICE BEAM

(b) **STANCHION AND LATTICE FRAME**

FIXED BASE

ROOF TRUSS

GANTRY GIRDER ON BRACKET

OVERHEAD TRAVELLING CRANE

(c) **STANCHION AND TRUSS FRAME WITH LIGHT CRANE**

FIXED BASE

ROOF TRUSS

LATTICED (OR BATTENED) STANCHION

E.O.T. CRANE

GANTRY GIRDER

(d) **LATTICE STANCHION AND TRUSS FRAME WITH HEAVY CRANE**

Figure 6.2.2 Single storey frame building

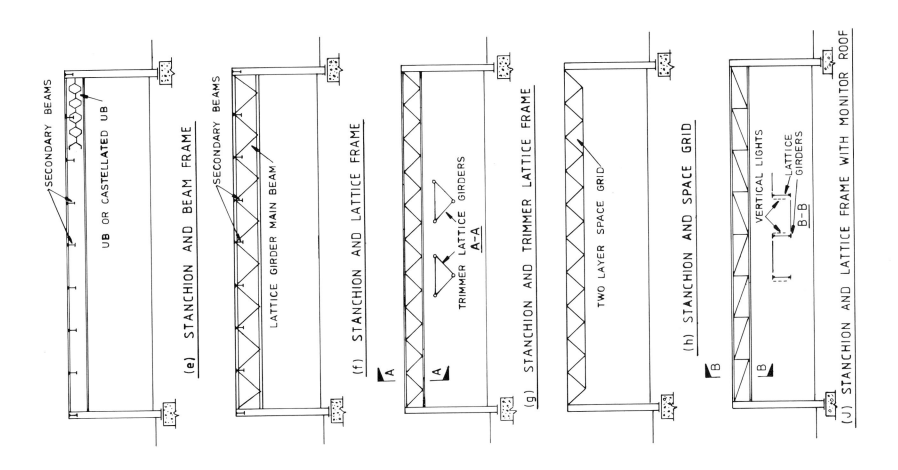

(e) STANCHION AND BEAM FRAME

SECONDARY BEAMS

UB OR CASTELLATED UB

(f) STANCHION AND LATTICE FRAME

SECONDARY BEAMS

LATTICE GIRDER MAIN BEAM

(g) STANCHION AND TRIMMER LATTICE FRAME

TRIMMER LATTICE GIRDERS

A-A

(h) STANCHION AND SPACE GRID

TWO LAYER SPACE GRID

(J) STANCHION AND LATTICE FRAME WITH MONITOR ROOF

VERTICAL LIGHTS

LATTICE GIRDERS

B-B

Figure 6.2.2 *Contd*

6.3 PORTAL FRAME BUILDINGS

Steel portal frames are the most common and are a particular form of single storey construction. They became popular from the 1950s and are particularly efficient in steel being able to make use of the plastic method of rigid design which enables sections of minimum weight to be used. Frame spacings of 4.5 m, 6.0 m and 7.5 m with roof pitch typically 1:10, 2:10 and 3:10 are common. Portal frames provide large clear floor areas offering maximum adaptability of the space inside the building. They are easily capable of being extended in the future and if known at the design stage built-in provision can be made. Multiple bays are possible. Variable eaves heights and spans can be achieved in the same building and selected internal columns can be deleted where required by the use of valley beams. Portal frames can be designed to accommodate overhead travelling cranes typically up to 10 tonnes capacity without use of compound stanchions.

Normally, wind loads on the gable ends are transferred via roof and side bracing systems within the end bays of the building to the foundations. The gable stanchions also provide fixings for the gable sheeting rails, which in turn support the cladding. Cold rolled section sheeting rails and purlins are usual, but alternatively hot rolled steel angle sections are suitable. Various proprietary systems are available using channel or zed sections. The sleeved system is popular whereby purlins extend over one bay between portal frames, but are made continuous over intermediate portals by a short sleeve of similar section. The systems often offer a range of fitments including rafter cleats, sag rods, rafter restraints, eaves beams, etc.

Main frame members are normally of universal beams with universal columns sometimes being used for the stanchions only. Tapered haunches (formed from cuttings of rafter section) are often introduced to strengthen the rafters at eaves, especially where a plastic design analysis has been used. Either pinned or fixed bases may be used. Main frames of tapering fabricated section are used by some fabricators, some of whom offer their own ranges of standard portal designs.

Bracing is essential for the overall stability of the structure especially during erection. Different arrangements to those illustrated may be necessary to accommodate door or window openings. It is important to provide restraint against buckling of rafters in the eaves region, this usually being supplied by an eaves beam together with diagonal stays connected to the purlins. Wind uplift forces often exceed the dead weight of portal frame buildings due to low roof pitch and light weight, such that holding down bolts must be supplied with bottom anchorage. Reversal of bending moments may also occur at eaves connections.

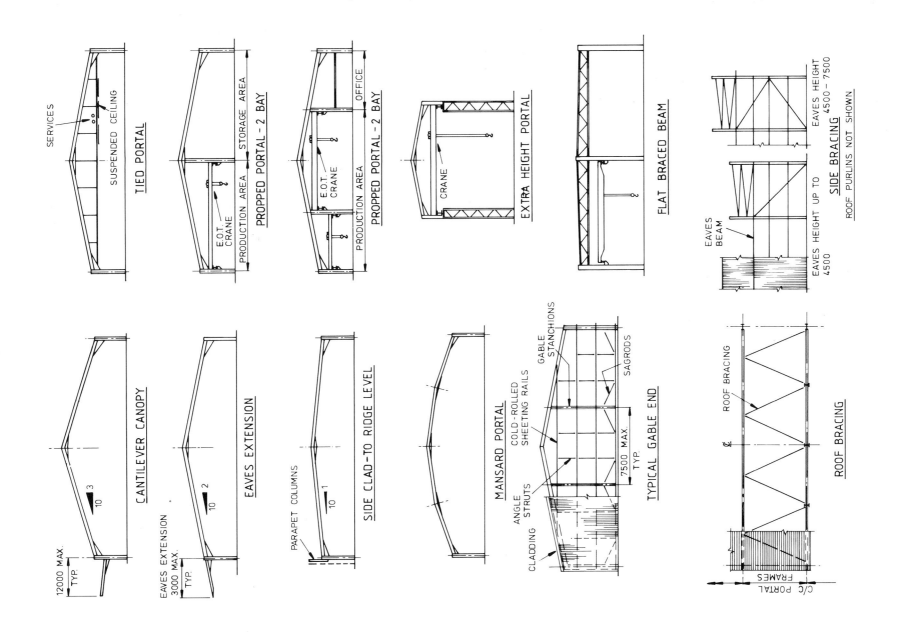

Figure 6.3.1 Portal frame buildings

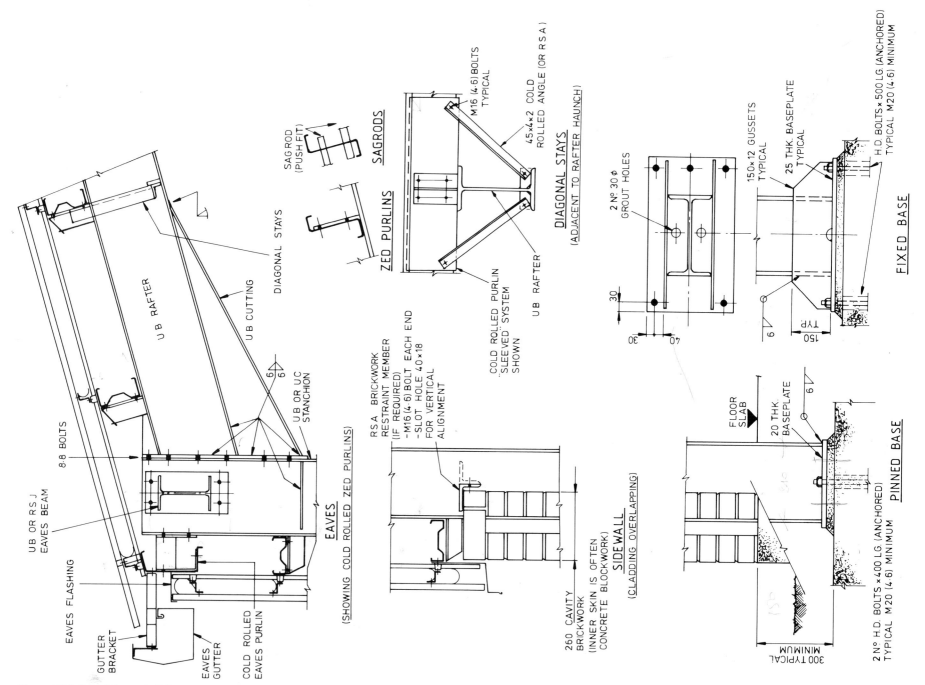

Figure 6.3.2 Portal frame buildings

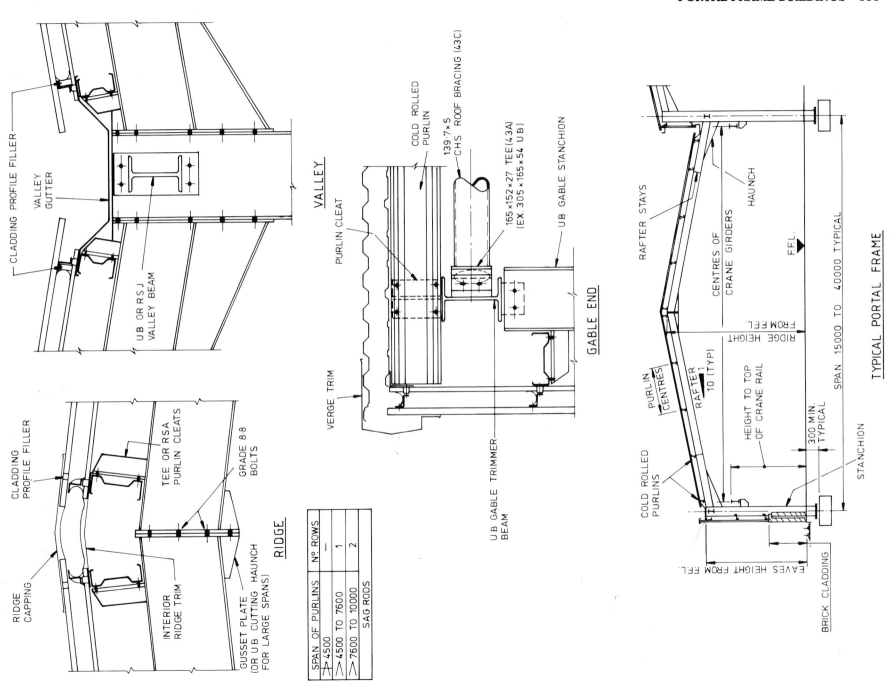

SPAN OF PURLINS	N° ROWS
⇗4500	—
⇗4500 TO 7600	1
⇗7600 TO 10000	2

SAG RODS

Figure 6.3.2 *Contd*

6.4 VESSEL SUPPORT STRUCTURE

The structure supports a carbon dioxide vessel weighing 12 tonnes 1.9 m diameter $\times$ 5.2 m long, approximately 3.1 m above ground level. It is typical of small supporting steelwork within industrial complexes and was installed inside a building. It comprises a main frame with four columns and beams made as one welded fabrication with rigid connections supporting the vessel cradle supplied by others. Access platforms are provided at two levels below and above the vessel with hooped access ladders.

Drawing notes

(1) All steel to be BS 4360:1986 grade 43A uos.
(2) All bolts to be black bolts grade 4.6. To be M16 diameter uos.
(3) All welds to be fillet welds size 5 mm uos continuous on both sides of all joints.
(4) Protective treatment all at workshop:
     Grit blast 2nd quality and zinc rich epoxy prefabrication primer.
     2 coats zinc rich epoxy paint after fabrication.
     Total nominal dry film thickness 150 microns.
     [System SD5 to BS 5493]

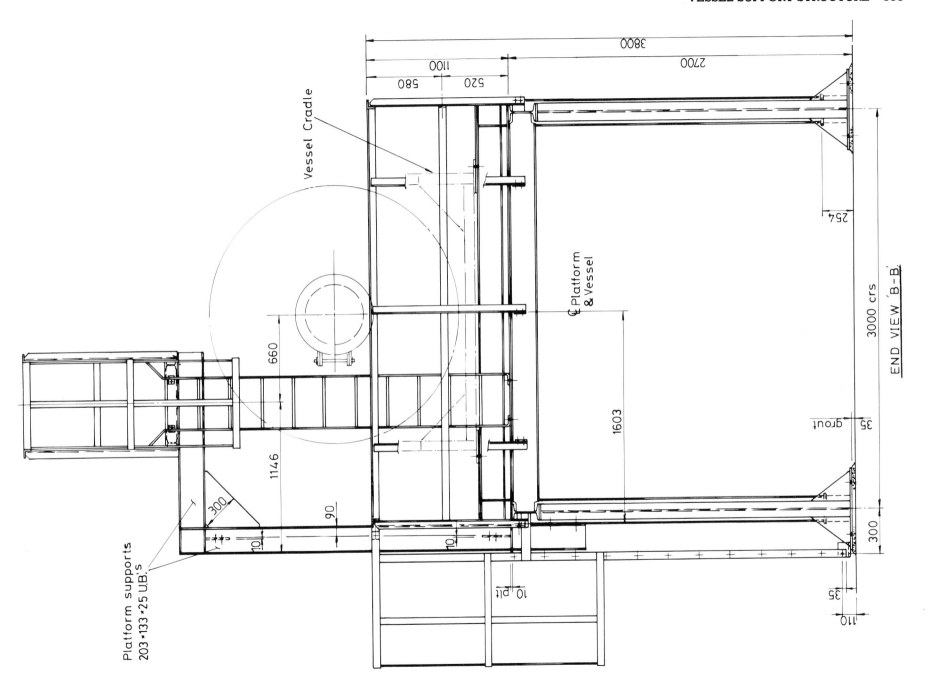

Figure 6.4.1 Vessel support structure

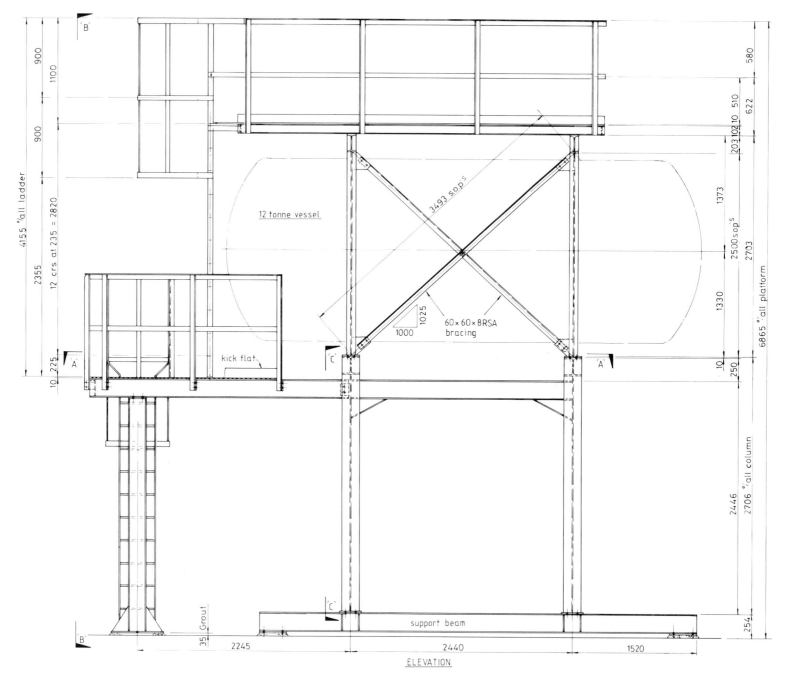

ELEVATION

Figure 6.4.2 Vessel support structure

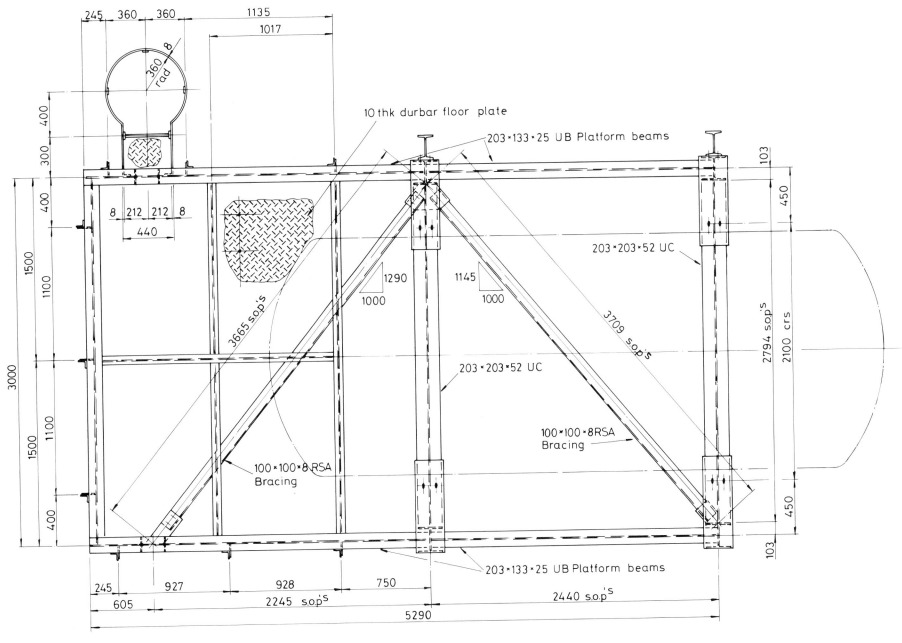

Figure 6.4.3 Vessel support structure

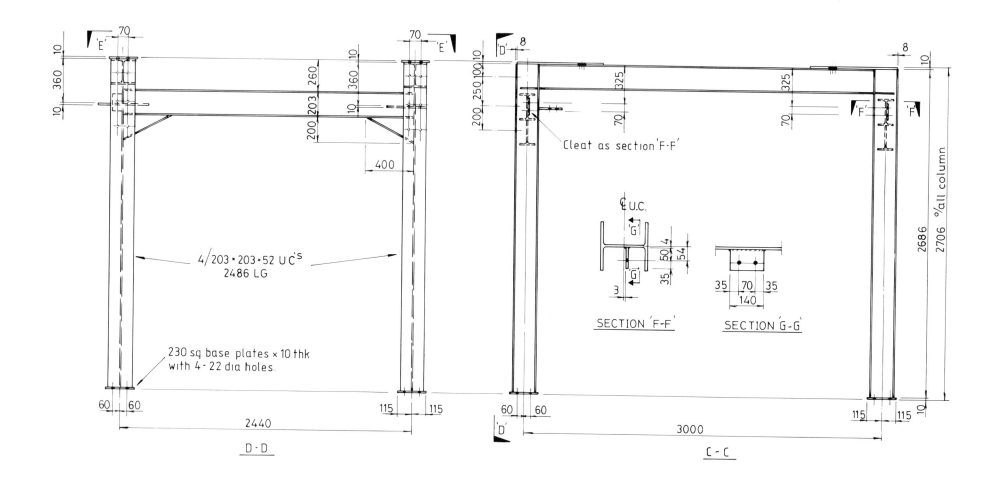

70
'E'
10
360
10

70   'E'
10
360
10

200   203   260
10
400

4/203×203×52 UC's
2486 LG

230 sq base plates × 10 thk
with 4-22 dia holes.

60   60
2440

115   115

D·D

8
'D'
200   250   100   10
325
70

Cleat as section 'F·F'

⊄ U.C.
'G'
4
50
54
35
3
'G'

SECTION 'F-F'

35   70   35
140

SECTION 'G-G'

325
70

'F'   'F'

2686   2706   %/all column

115   115
3000
10

C - C

60   60
'D'

1- SUPPORT PLATFORM REQ'D AS DRAWN

Figure 6.4.3 *Contd*

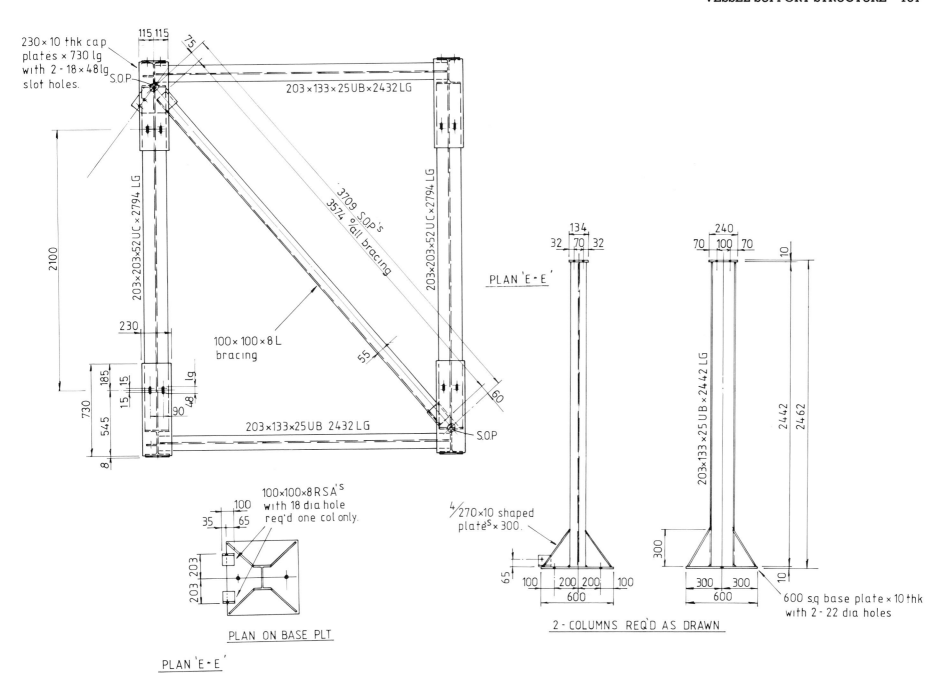

Figure 6.4.4 Vessel support structure

6.5 ROOF OVER RESERVOIR

The roof provides a protective covering over a fresh water reservoir with a span of about 19.5 m which is clad with profiled steel sheeting. It comprises pitched universal beam rafters which are tied at eaves level with RSA ties because the reservoir edge walls are not capable of resisting outward horizontal thrust. The ties are supported from the ridge at mid-length to prevent sagging. Roof plan bracing is supplied within one internal bay to ensure longitudinal stability of the roof.

Drawing notes

(1)   All steel to be BS 4360:1986 grade 43A uos.
(2)   All bolts to be black bolts grade 4.6 uos.
       To be M16 diameter uos.
(3)   All welds to be fillet welds size 6 mm uos continuous on both sides of all joints.
(4)   Protective treatment:
       Grit blast 2nd quality and zinc rich epoxy prefabrication primer.
       One coat zinc rich epoxy paint at workshop.
       One coat zinc rich epoxy paint at site after erection.
       Total nominal dry film thickness 150 microns.
       [System SD5 to BS 5493]

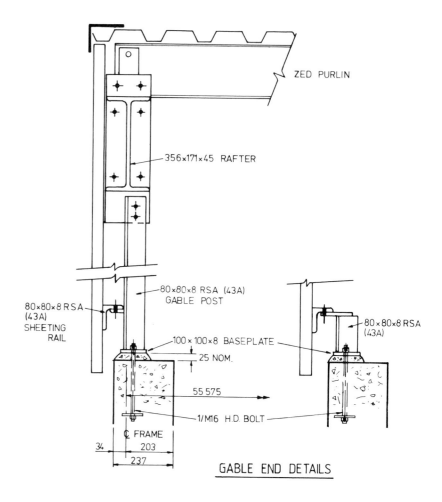

Figure 6.5.1 Roof over reservoir

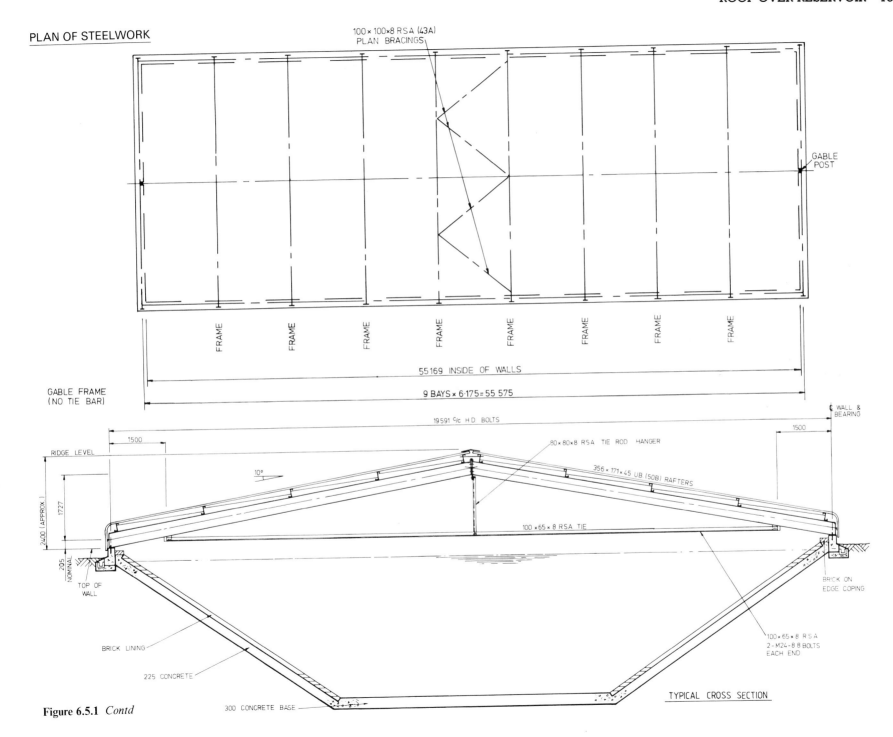

PLAN OF STEELWORK

100 × 100×8 R S A (43A)
PLAN BRACINGS

GABLE
POST

FRAME FRAME FRAME FRAME FRAME FRAME FRAME FRAME

55169 INSIDE OF WALLS

GABLE FRAME
(NO TIE BAR)

9 BAYS × 6·175 = 55 575

19 591 c/c H D BOLTS

WALL &
BEARING

1500

1500

RIDGE LEVEL

80 × 80 × 8 R S A TIE ROD HANGER

10°

356 × 171·45 UB (50B) RAFTERS

2400 (APPROX)

1727

100 × 65 × 8 RSA TIE

205
NOMINAL

TOP OF
WALL

BRICK ON
EDGE COPING

BRICK LINING

225 CONCRETE

100 × 65 × 8 R S A
2 - M24 - 8.8 BOLTS
EACH END

300 CONCRETE BASE

TYPICAL CROSS SECTION

Figure 6.5.1 *Contd*

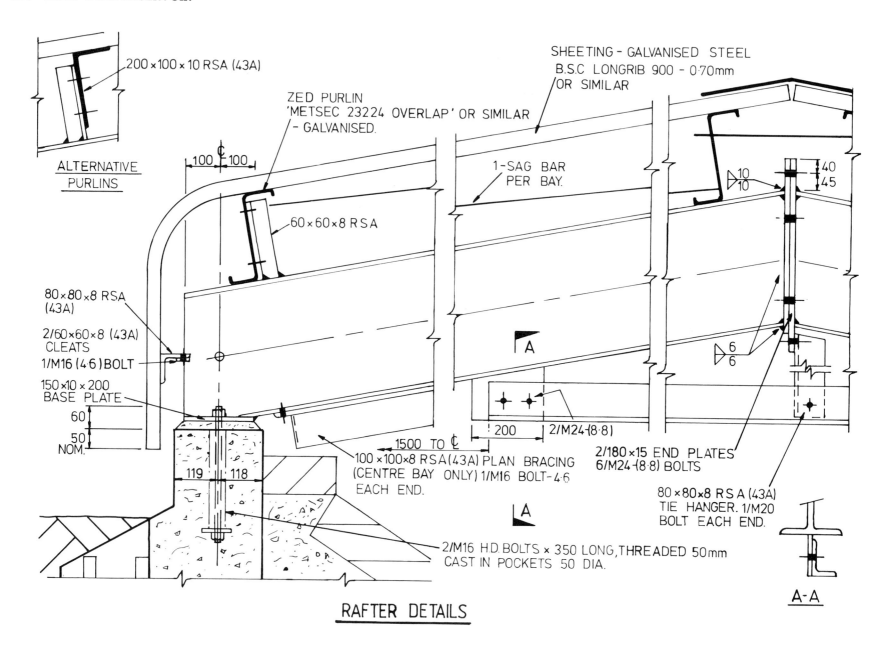

ALTERNATIVE PURLINS

200×100×10 RSA (43A)

ZED PURLIN
'METSEC 23224 OVERLAP' OR SIMILAR
- GALVANISED.

SHEETING - GALVANISED STEEL
B.S.C LONGRIB 900 - 0·70mm
OR SIMILAR

100 ℄ 100

1-SAG BAR
PER BAY.

60×60×8 RSA

80×80×8 RSA
(43A)

2/60×60×8 (43A)
CLEATS

1/M16 (4·6) BOLT

150×10×200
BASE PLATE

60
50
NOM.

119   118

1500 TO ℄

100×100×8 RSA(43A) PLAN BRACING
(CENTRE BAY ONLY) 1/M16 BOLT-4·6
EACH END.

200   2/M24-(8·8)

10
10

40
45

6
6

2/180×15 END PLATES
6/M24-(8·8) BOLTS

80×80×8 RSA (43A)
TIE HANGER. 1/M20
BOLT EACH END.

2/M16 H.D.BOLTS × 350 LONG,THREADED 50mm
CAST IN POCKETS 50 DIA.

A

A

A-A

RAFTER DETAILS

Figure 6.5.2 Roof over reservoir

6.6 TOWER

The tower is 55 m high and supports electrical equipment within an electricity power generating station in India. It was fabricated in the UK and transported piecemeal by ship in containers. The major consideration in the design of tower structures is wind loading due to the height above ground and comparatively light weight of the equipment carried. Open braced structures are usual for towers so as to offer minimal wind resistance. Either hollow sections or rolled angles would have been suitable and although the former have an advantage in providing for smooth air flow and thus less wind resistance, the latter were chosen to simplify the connections. Use of bolted connections using gusset plates meant that all members could be economically fabricated using NC saw/drilling equipment.

Notes

(1)   All steel to be to BS 4360:1986 grade 43A uos.
(2)   All bolts to be grade 4.6. To be M24 diameter uos.

TYPICAL   WORKSHOP   DETAIL

Figure 6.6.1 Tower

SAFETY CAGE MATERIAL
HOOPS   50 x 10 FLAT (43A)
STRAPS   50 x 10 FLAT (43A)
ALL FIXING   BOLTS
M12 x 40 LG   CUP   HEAD

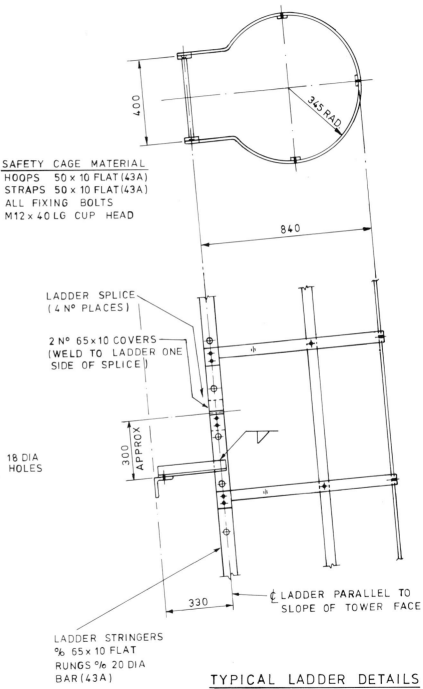

LADDER SPLICE
(4 Nº PLACES)

2 Nº 65 x 10 COVERS
(WELD TO LADDER ONE
SIDE OF SPLICE)

₡ LADDER PARALLEL TO
SLOPE OF TOWER FACE

LADDER STRINGERS
% 65 x 10 FLAT
RUNGS % 20 DIA
BAR (43A)

TYPICAL   LADDER   DETAILS

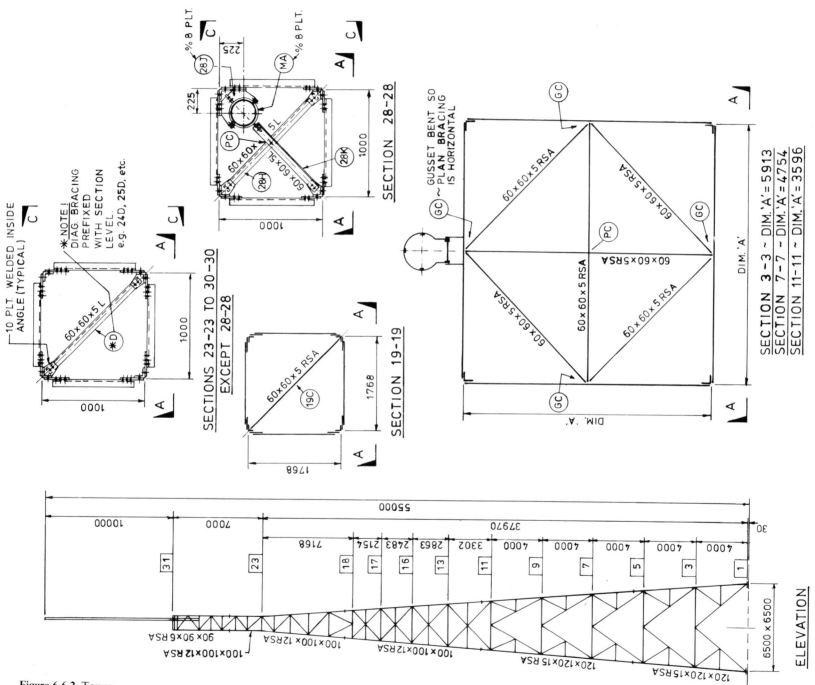

Figure 6.6.2 Tower

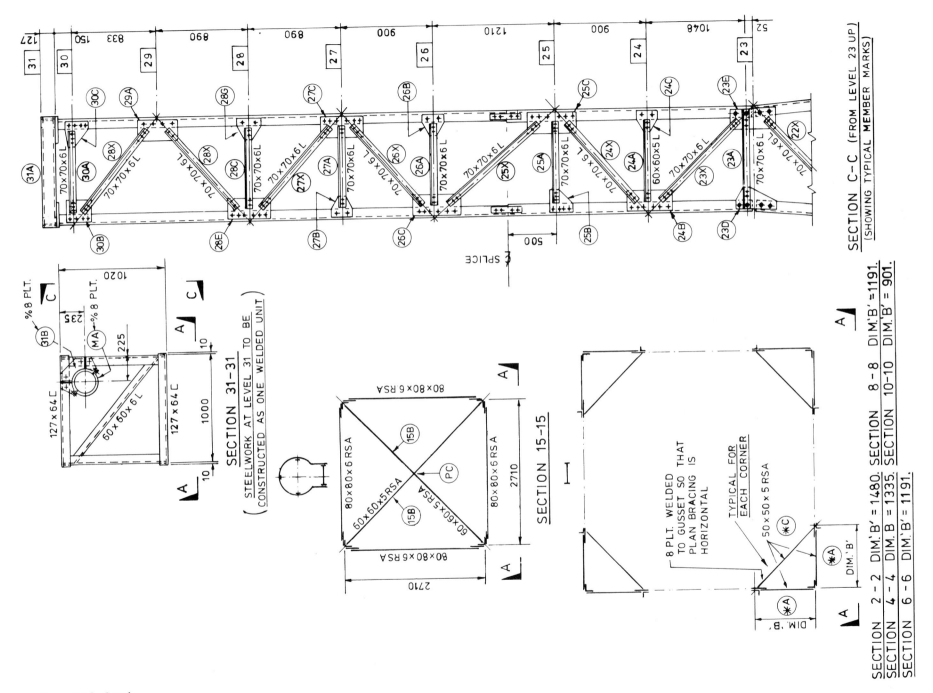

SECTION C–C (FROM LEVEL 23 UP)
(SHOWING TYPICAL MEMBER MARKS)

SECTION 31–31
(STEELWORK AT LEVEL 31 TO BE
CONSTRUCTED AS ONE WELDED UNIT)

SECTION 15–15

SECTION 2 – 2   DIM.'B' = 1480.   SECTION   8 – 8   DIM.'B' =1191.
SECTION 4 – 4   DIM.'B = 1335.   SECTION 10–10   DIM.'B' = 901.
SECTION 6 – 6   DIM.'B' = 1191.

Figure 6.6.2 Contd

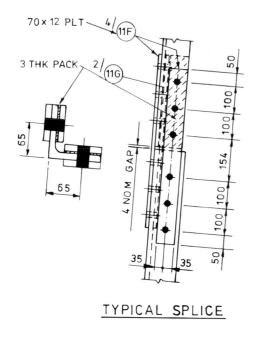

70 × 12 PLT — 4/(11F)

3 THK PACK — 2/(11G)

65

65

4 NOM GAP

50
100 100
100
154
100
100
50
35   35

TYPICAL SPLICE

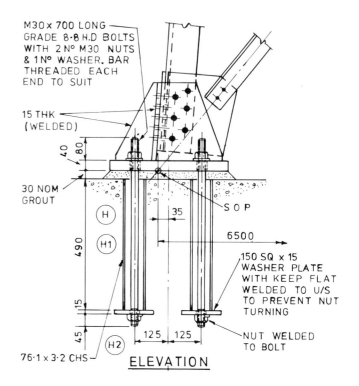

M30 × 700 LONG GRADE 8·8 H.D BOLTS WITH 2 N° M30 NUTS & 1 N° WASHER. BAR THREADED EACH END TO SUIT

15 THK (WELDED)

30 NOM GROUT

40
80
490
15
45

(H)
(H1)
(H2)

35

6500

S O P

150 SQ × 15 WASHER PLATE WITH KEEP FLAT WELDED TO U/S TO PREVENT NUT TURNING

125   125

NUT WELDED TO BOLT

76·1 × 3·2 CHS

ELEVATION

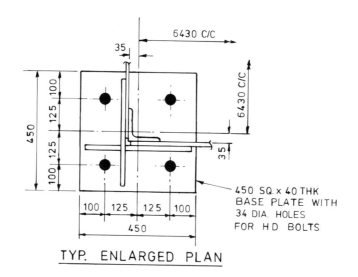

6430 C/C

35

6430 C/C

450
125 100
125
100 125
100

100   125   125   100
450

35

450 SQ. × 40 THK BASE PLATE WITH 34 DIA. HOLES FOR H D BOLTS

TYP. ENLARGED PLAN

Figure 6.6.3 Tower

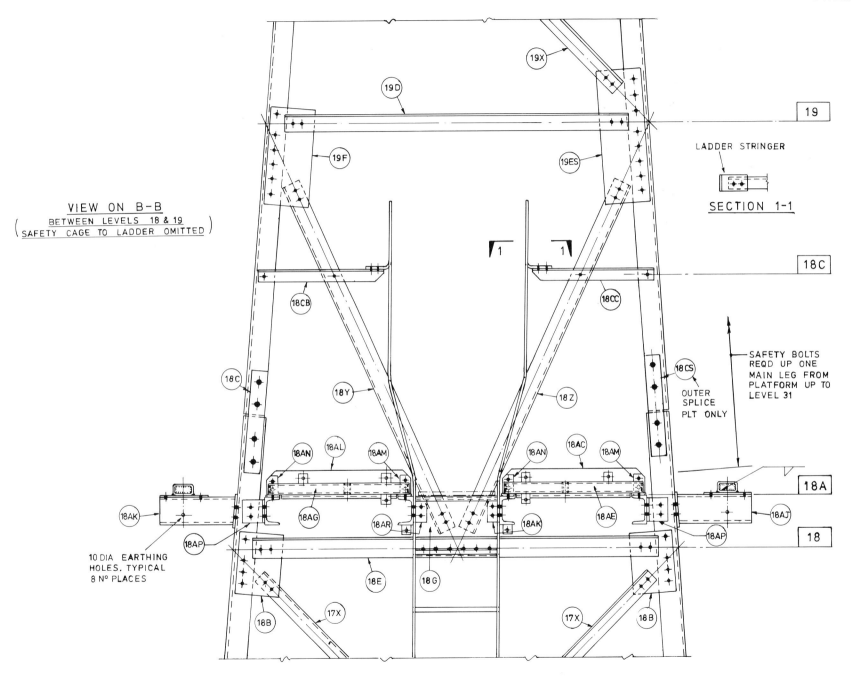

VIEW ON B-B
(BETWEEN LEVELS 18 & 19)
(SAFETY CAGE TO LADDER OMITTED)

LADDER STRINGER

SECTION 1-1

19

18C

18A

18

SAFETY BOLTS
REQD UP ONE
MAIN LEG FROM
PLATFORM UP TO
LEVEL 31

OUTER
SPLICE
PLT ONLY

10 DIA EARTHING
HOLES. TYPICAL
8 N° PLACES

Figure 6.6.3 *Contd*

6.7 BRIDGES

Several developments since the late 1970s have improved the status for steel in bridges increasing its market share over concrete structures in a number of countries of the world. Developments include:

(a)   Fabricators have improved their efficiency by use of automation.
(b)   Stability of steel prices with wider availability in many countries by opening of steel plants.
(c)   Use of mobile cranes to erect large pre-assembled components quickly, thus reducing numbers of mid-air joints.
(d)   Composite construction economises on materials.
(e)   Permanent formwork or precasting for slabs.
(f)   Improved protection systems using fewer paint coats having longer life.
(g)   Use of unpainted weathering steel for inaccessible bridges.
(h)   Use of site welded or HSFG bolted joints to achieve continuous spans.
(j)   Better education in steel design.

For multiple short (up to 30 m) and medium spans (30 m to 150 m) continuity is common with welded or HSFG bolted site joints to the main members. Articulation between deck and substructures is generally provided using sliding or pinned bearings mounted on vertical piers often of concrete but occasionally steel. Constant depth main girders are usual, with fabricated precamber to counteract deflection. Curved soffits are sometimes used (as shown in figure 6.7.1).

Curved bridges are often formed using straight fabricated chords with change of direction at site splices. Composite *deck type* cross sections are usual for highway bridges as shown in figure 6.7.2 and suit the width of modern roads except where construction depth is very restricted when half-through girders are used, especially for railway bridges as shown in figure 6.7.3. Multiple rolled sections are used for short spans with plate girders being used when the span exceeds about 25 to 30 m. Intermediate lateral bracings are provided for stability. Sometimes they are proportioned to assist in transverse distribution of live load, but practices vary between different countries. Box girders as shown in figure 6.7.3 are also used and open top boxes 'bathtubs' are extensively used in North America. Problems can arise during construction due to distortion and twisting of open top boxes prior to the rigidifying effect of the concrete slab being realised and temporary bracings are thus essential.

Most early composite bridges used *in situ* slabs cast on removable formwork supported from the steelwork. Recently the high costs of timber and site labour have encouraged permanent formwork. Various types are in use including profiled steel sheeting (especially in the USA), glass reinforced plastic (grp), glass reinforced concrete (grc) and part depth concrete planks. The 'OMNIA' type of precast unit is being used (see figure 5.11) which incorporates a welded lattice truss to provide temporary capacity to span up to about 3.5 m between steel flanges, whose lower chord is cast in. Extra reinforcement is incorporated supplemented by further continuous rebars at the 'vee' joints to resist live loads. Detailing of the slab needs to be carefully done to avoid congestion of reinforcement and allow proper compaction of concrete.

For *footbridges* steel provides a good solution because the entire cross section including parapets can be erected in one piece. Cross sections are shown in figure 6.7.4. Economic solutions use half-through lattice or Vierendeel girders with members of rolled hollow section and deck plate with factory applied epoxy-type non-slip surfacing 6 mm or less in thickness. Columns, staircases and ramps are also commonly of steel using hollow sections. For urban areas the half-through section achieves minimum length approach stairs or ramps. Further space can be saved by using *stepped ramps* which achieve an average slope of 1 in 6 compared with 1 in 10 for sloping ramps.

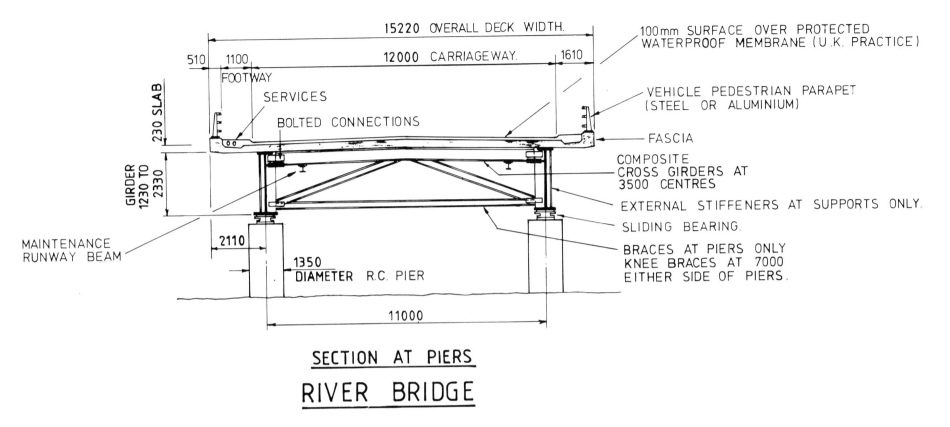

15220 OVERALL DECK WIDTH.

12000 CARRIAGEWAY.

510 1100 1610

FOOTWAY

230 SLAB

GIRDER 1230 TO 2330

SERVICES

BOLTED CONNECTIONS

100mm SURFACE OVER PROTECTED WATERPROOF MEMBRANE (U.K. PRACTICE)

VEHICLE PEDESTRIAN PARAPET (STEEL OR ALUMINIUM)

FASCIA

COMPOSITE CROSS GIRDERS AT 3500 CENTRES

EXTERNAL STIFFENERS AT SUPPORTS ONLY.

SLIDING BEARING.

BRACES AT PIERS ONLY KNEE BRACES AT 7000 EITHER SIDE OF PIERS.

MAINTENANCE RUNWAY BEAM

2110

1350 DIAMETER R.C. PIER

11000

SECTION AT PIERS
RIVER BRIDGE

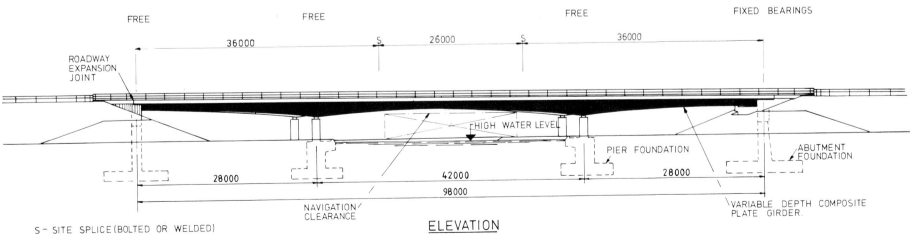

FREE FREE FREE FIXED BEARINGS

36000 S 26000 S 36000

ROADWAY EXPANSION JOINT

HIGH WATER LEVEL

PIER FOUNDATION

ABUTMENT FOUNDATION.

28000 42000 28000

98000

NAVIGATION CLEARANCE

VARIABLE DEPTH COMPOSITE PLATE GIRDER.

S - SITE SPLICE (BOLTED OR WELDED)

ELEVATION

Figure 6.7.1 Bridges

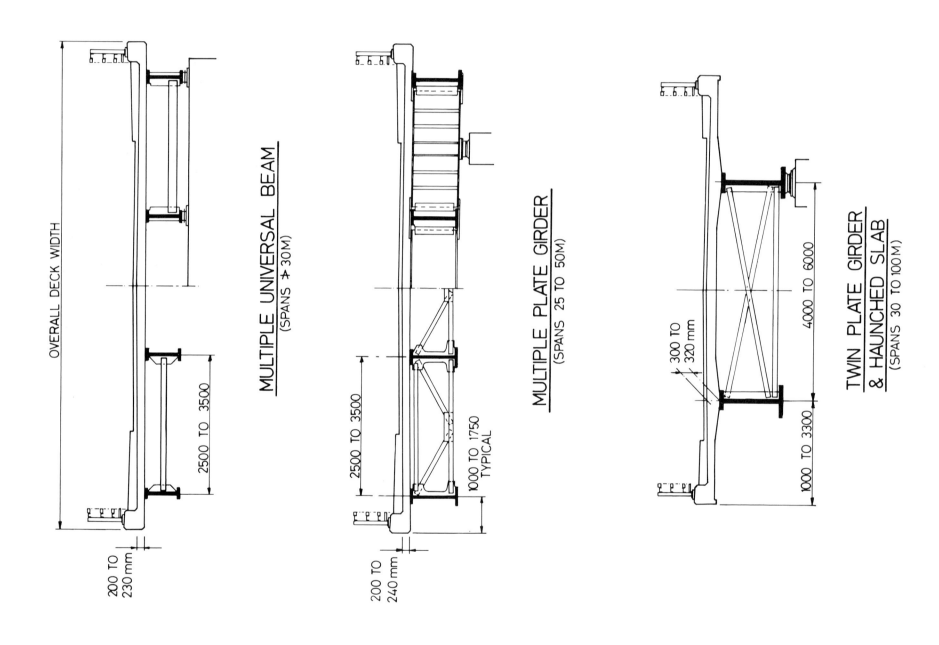

MULTIPLE UNIVERSAL BEAM
(SPANS ≯ 30M)

OVERALL DECK WIDTH

2500 TO 3500

200 TO 230 mm

MULTIPLE PLATE GIRDER
(SPANS 25 TO 50M)

2500 TO 3500

1000 TO 1750 TYPICAL

200 TO 240 mm

TWIN PLATE GIRDER
& HAUNCHED SLAB
(SPANS 30 TO 100M)

4000 TO 6000

1000 TO 3300

300 TO 320 mm

Figure 6.7.2 Bridges

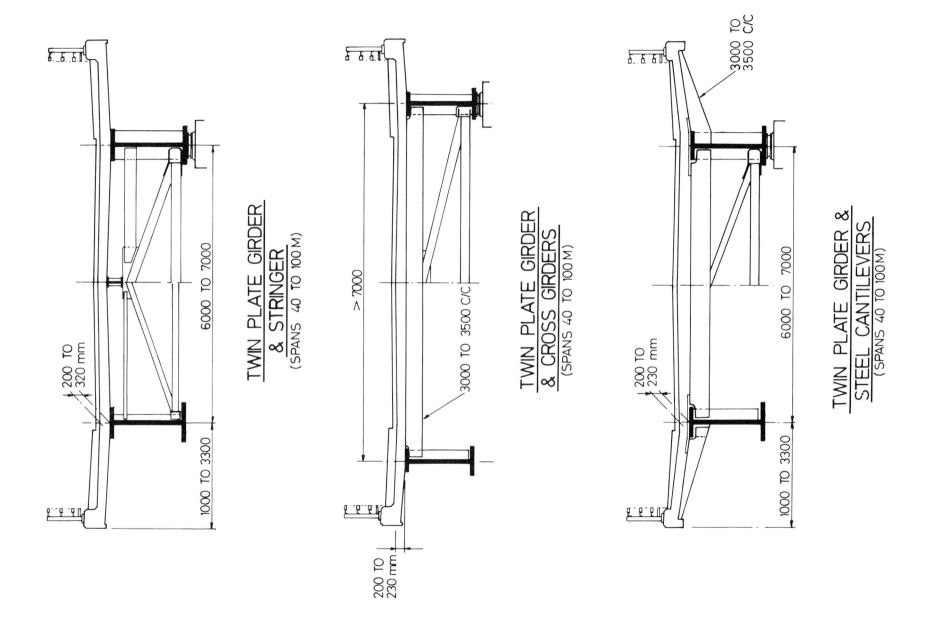

TWIN PLATE GIRDER & STRINGER
(SPANS 40 TO 100 M)

TWIN PLATE GIRDER & CROSS GIRDERS
(SPANS 40 TO 100 M)

TWIN PLATE GIRDER & STEEL CANTILEVERS
(SPANS 40 TO 100 M)

Figure 6.7.2 *Contd*

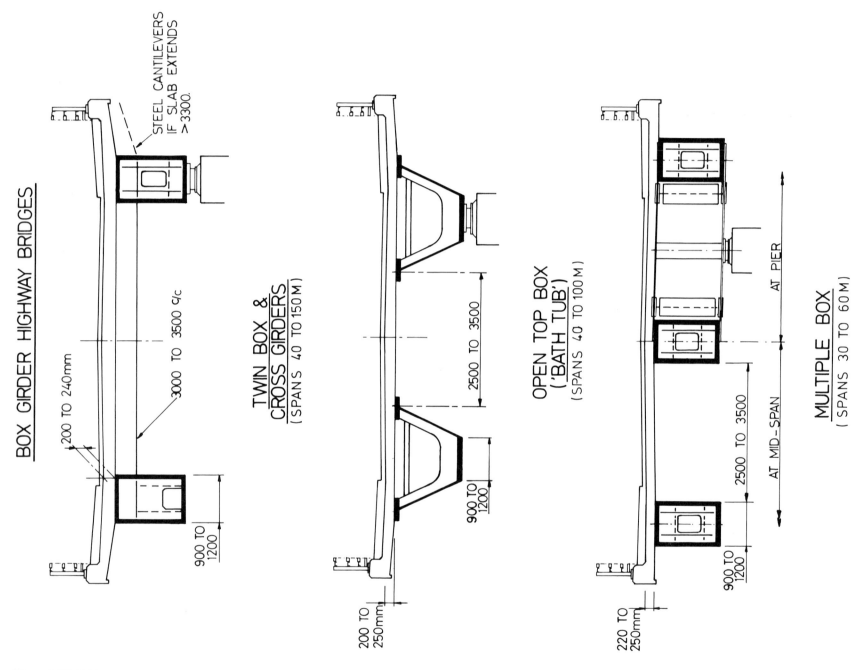

Figure 6.7.3 Bridges

BOX GIRDER HIGHWAY BRIDGES

STEEL CANTILEVERS IF SLAB EXTENDS >3300.

200 TO 240mm

3000 TO 3500 c/c

900 TO 1200

TWIN BOX & CROSS GIRDERS
(SPANS 40 TO 150 M)

2500 TO 3500

900 TO 1200

200 TO 250mm

OPEN TOP BOX ('BATH TUB')
(SPANS 40 TO 100 M)

AT PIER

AT MID-SPAN

2500 TO 3500

900 TO 1200

220 TO 250mm

MULTIPLE BOX
(SPANS 30 TO 60 M)

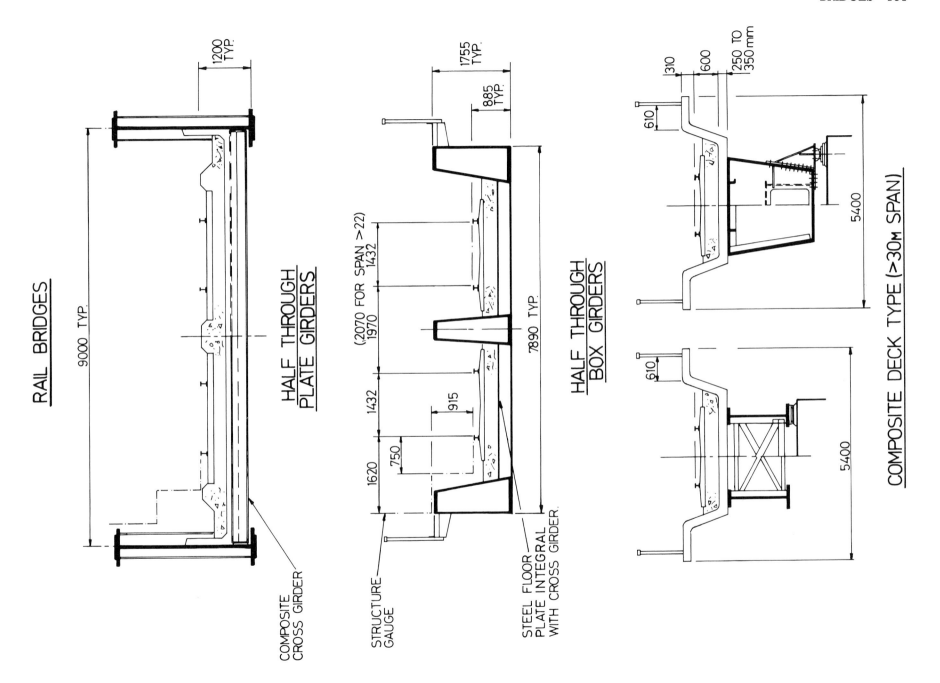

RAIL BRIDGES

1200 TYP.

9000 TYP.

HALF THROUGH
PLATE GIRDERS

COMPOSITE
CROSS GIRDER

1755 TYP.

885 TYP.

(2070 FOR SPAN >22)
1970

1432

1432

1620

915

750

7890 TYP.

HALF THROUGH
BOX GIRDERS

STRUCTURE
GAUGE

STEEL FLOOR
PLATE INTEGRAL
WITH CROSS GIRDER.

310

600

250 TO
350 mm

610

5400

610

5400

COMPOSITE DECK TYPE (>30M SPAN)

Figure 6.7.3 *Contd*

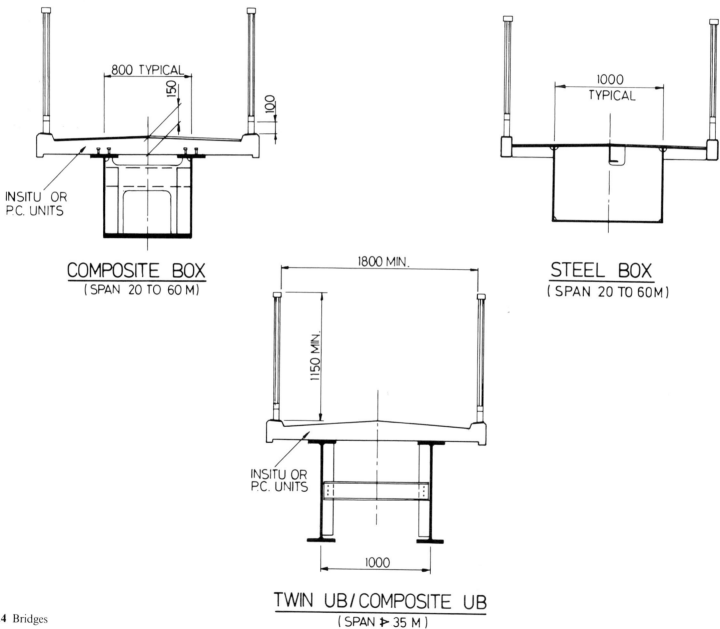

800 TYPICAL

150

100

INSITU OR
P.C. UNITS

COMPOSITE BOX
(SPAN 20 TO 60 M)

1000
TYPICAL

STEEL BOX
(SPAN 20 TO 60M)

1800 MIN.

1150 MIN.

INSITU OR
P.C. UNITS

1000

TWIN UB/COMPOSITE UB
(SPAN ⊅ 35 M)

Figure 6.7.4 Bridges

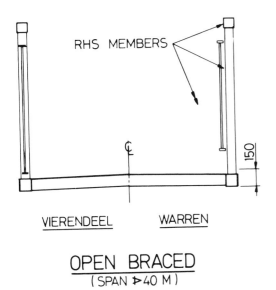

RHS MEMBERS

150

VIERENDEEL   WARREN

OPEN BRACED
(SPAN ⊳ 40 M)

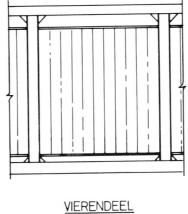

VIERENDEEL

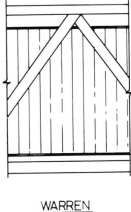

WARREN

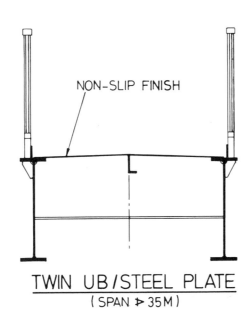

NON-SLIP FINISH

TWIN UB / STEEL PLATE
(SPAN ⊳ 35 M)

Figure 6.7.4 *Contd*

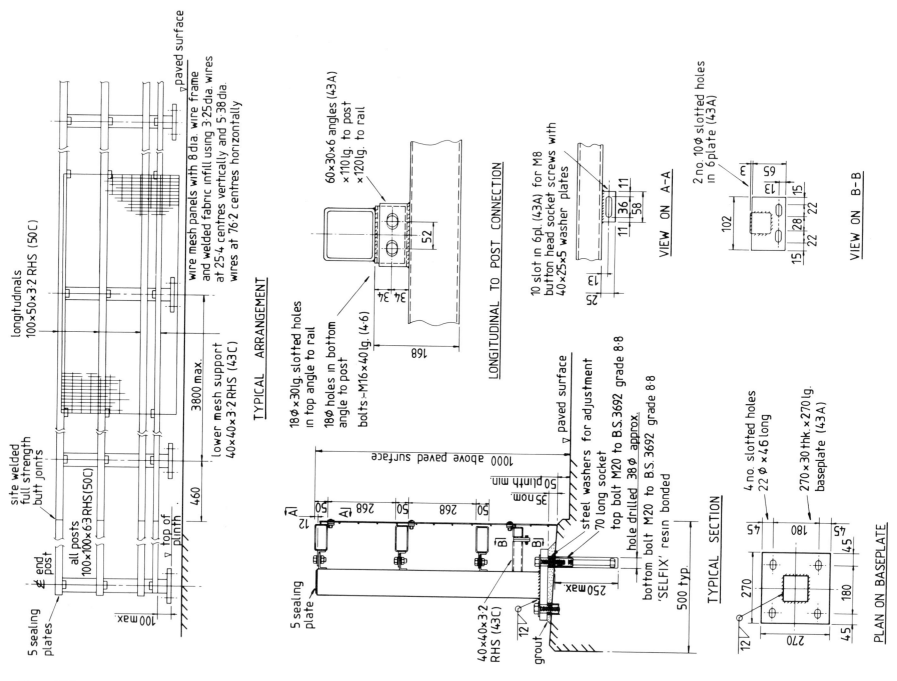

TYPICAL ARRANGEMENT

longitudinals 100×50×3·2 RHS (50C)

wire mesh panels with 8 dia. wire frame and welded fabric infill using 3·25 dia. wires at 25·4 centres vertically and 5·38 dia. wires at 76·2 centres horizontally

▽ paved surface

site welded full strength butt joints

all posts 100×100×6·3 RHS (50C)

5 sealing plates

100 max.

460

3800 max.

lower mesh support 40×40×3·2 RHS (43C)

▽ top of plinth

€ end post

LONGITUDINAL TO POST CONNECTION

60×30×6 angles (43A) ×110 lg. to post ×120 lg. to rail

52

34 34

168

18ø×30 lg. slotted holes in top angle to rail

18ø holes in bottom angle to post

bolts:- M16×40 lg. (4·6)

VIEW ON A–A

10 slot in 6 pl. (43A) for M8 button head socket screws with 40×25×5 washer plates

11 11

36

58

25

13

VIEW ON B–B

2 no. 10ø slotted holes in 6 plate (43A)

3

65

13 15

28 22

22 22

102

15

TYPICAL SECTION

▽ paved surface

1000 above paved surface

50 plinth min.

35 nom.

steel washers for adjustment

70 long socket

top bolt M20 to B.S. 3692 grade 8·8

hole drilled 38ø approx.

bottom bolt M20 to B.S. 3692 grade 8·8 'SELFIX' resin bonded

500 typ.

250 max.

A

A

50

268

50

50

268

50

12

B B

5 sealing plate

40×40×3·2 RHS (43C)

grout

12

PLAN ON BASEPLATE

4 no. slotted holes 22ø×46 long

270×30 thk.×270 lg. baseplate (43A)

45

180

45

270

180

45

270

180

45

12

Figure 6.7.5 Bridges

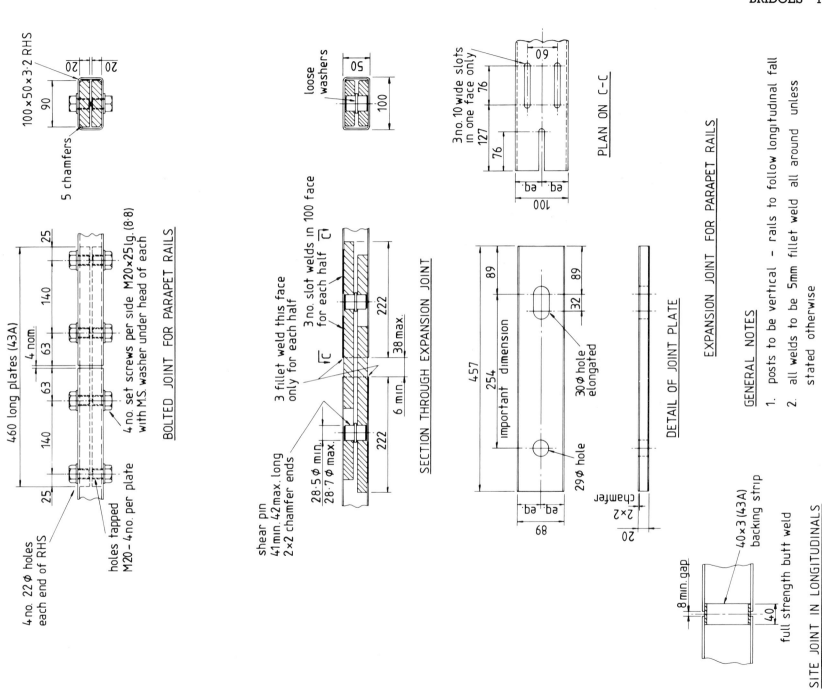

Figure 6.7.5 *Contd*

6.8 SINGLE SPAN HIGHWAY BRIDGE

The bridge carries a motorway across railway tracks with a clear span of 31.5 between r.c. abutments and an overall width of 35.02 m. It is suitable for dual three-lane carriageways, hard shoulders and central reserve. It can be adapted to suit different highway widths. Plan curvature of the motorway is accommodated by an increased deck width. Use of steel plate girders with permanent slab formwork allows rapid construction over the railway and would also be suitable across a river. Weathering steel is used to avoid future maintenance painting.

Composite plate girders at 3.08 m centers support the 255 mm thick deck slab and finishes. The edge girders are 1.6 m deep and carry the extra weight of the parapets which are solid reinforced concrete 'high containment' type. In other locations a lighter open steel parapet is more usual as shown in figure 6.7.5.

Inner girders are 1.3 m deep. They are shown fabricated in a single length, but in the UK special permission is required for movement of loads exceeding 27.4 m and this is normally only feasible if good road access is available from the fabrication works, or if rail transport is used. Alternative bolted or welded site splices are shown in figure 6.8.4. The minimum number of flange thickness changes are made, consistent with available plate lengths. This avoids the high costs of making full penetration butt welds. The girders are precambered in elevation so as to counteract dead load deflection and to follow the road geometry. For calculation of the deflection girder self weight and concrete slab is assumed carried by the girder alone, whilst finishes and parapets are taken by the composite section. It may be noted that a typical precamber for composite girders is about 0.25% to 0.5% of span.

Girders are fixed against longitudinal movement at one abutment and free to move at the other. Bearings are proprietary 'pot' or 'disc' type bearings comprising a rubber disc contained within a steel cylinder and piston arrangement. The rubber, being contained, is able to withstand high vertical loads whilst permitting rotation. The free abutment bearings incorporate ptfe (polytetra fluorethylane) stainless steel sliding surfaces to cater for thermal movements and concrete shrinkage. Composite steel channel trimmers occur at each abutment to restrain the girders during construction and to stiffen the slab ends. Within the span two lines of transverse channel bracings are provided for erection stability. All site connections are made using HSFG bolts. For erection the girders were placed in groups of up to three using a lifting beam as shown in figure 6.8.5. This is convenient where the erection period is limited by short railway occupations and was used to erect the prototype of the bridge described.

Drawing notes

(1) All steel to be weather resistant unpainted to BS 4360 grade WR50C uos.
(2) All bolts to be HSFG to BS 4395 Part 1. Chemical composition to ASTM A325 Type 3, Grade A, or equivalent weather resistant. To be M24 diameter uos.
(3) Intermediate stiffeners may be radial to camber.
(4) All welds to be fillet welds size 6 mm uos continuous on both sides of all joints.
(5) Butt welds – all transverse welds to flanges and webs to be full penetration welds.
(6) All welding electrodes shall be to BS 639 Sections 1 and 4. Welds shall possess similar weather resisting properties to the steel such that these are retained, including possible loss of thickness due to slow rusting. The design allows for loss of thickness of 2 mm on all exposed surfaces.
(7) Temporary lifting cleats may remain in position within slab.
(8) Temporary welds shall not occur within 25 mm of any flange edge.
(9) Complete trial erection of three adjacent plate girders shall be performed. During the trial erection the true relative levels of the steelwork shall be modelled.
(10) The exposed outer surfaces of web top flange and bottom flange including soffit to girders 1 and 12, together with all HSFG interfaces, shall be blast cleaned to 3rd quality BS 4232. All other surfaces shall be maintained free from contamination by concrete, mortar, asphalt, paint, oil, grease and any other undesirable contaminants.

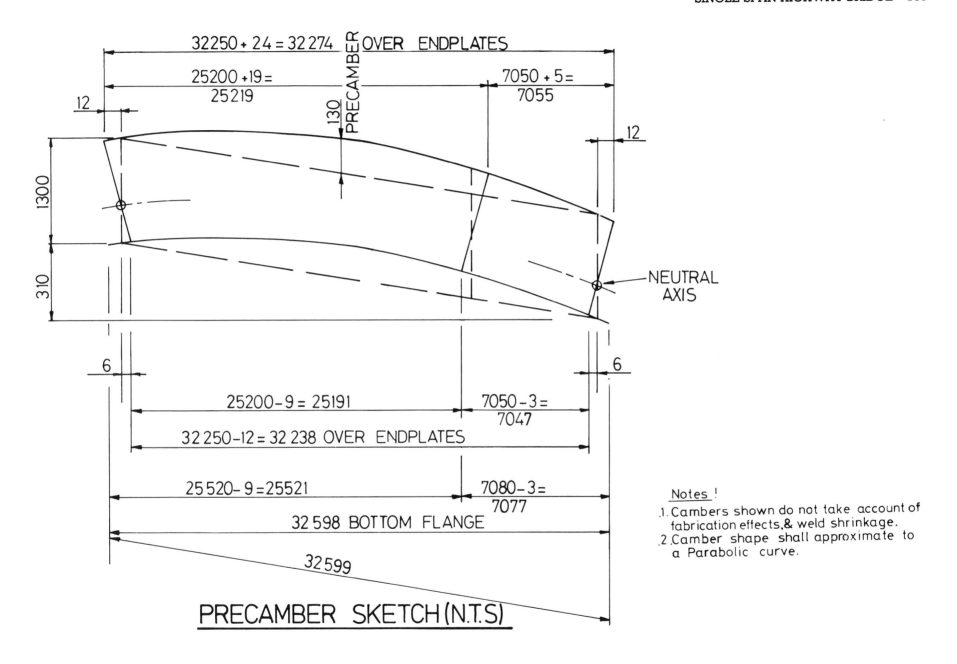

Figure 6.8.1 Single span highway bridge

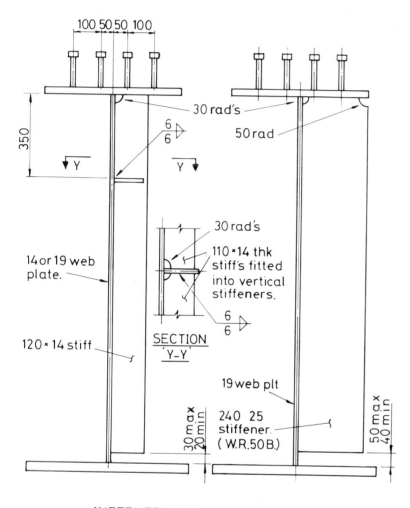

INTERMEDIATE STIFFENERS.

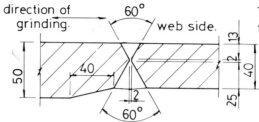

BOTTOM FLANGE BUTT WELD.

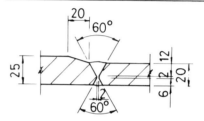

TOP FLANGE BUTT WELD.

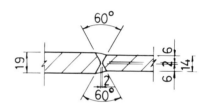

WEB BUTT WELD.

Precamber at mid-span Girders 2. to 11.	
Girder weight.	21
Slab. etc.	62
Finishes.	15
Shrinkage.	15
Final precamber.	15
Total.	128
Specified precamber.	130

Figure 6.8.2 Single span highway bridge

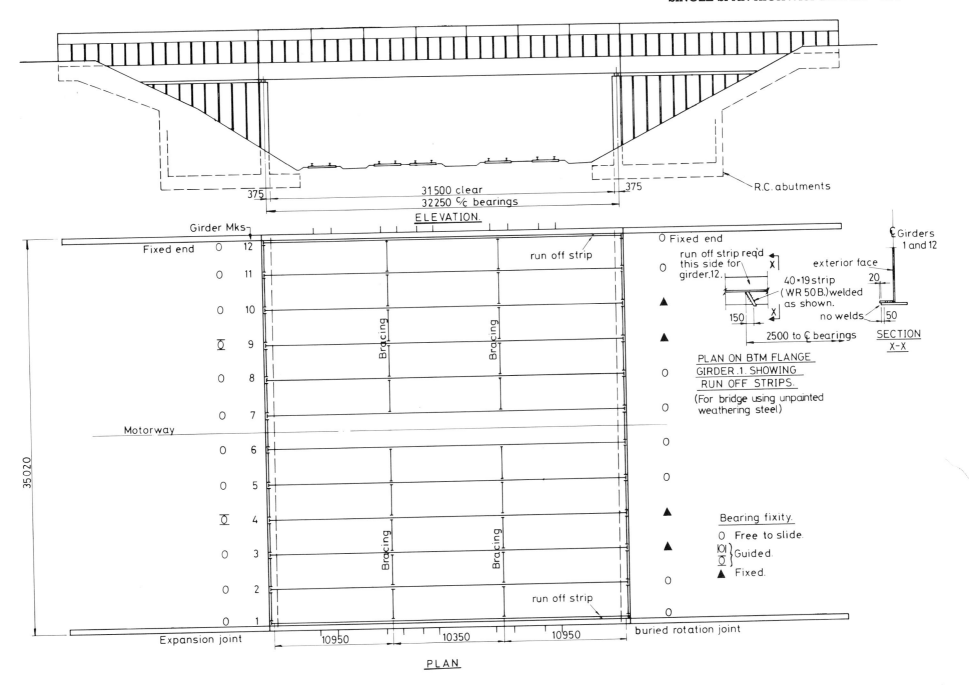

Figure 6.8.3 Single span highway bridge

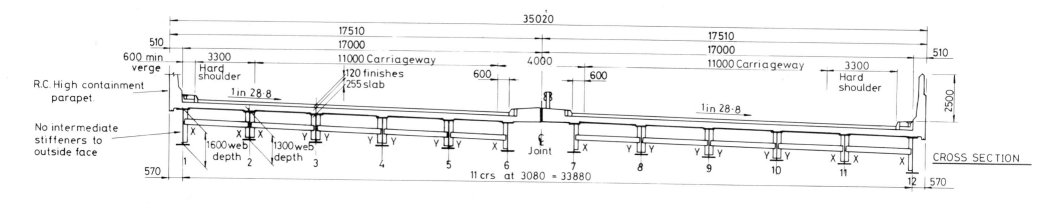

Figure 6.8.3 *Contd*

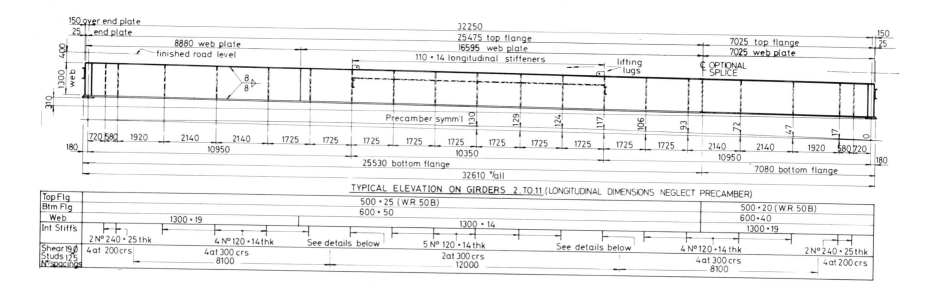

Figure 6.8.4 Single span highway bridge

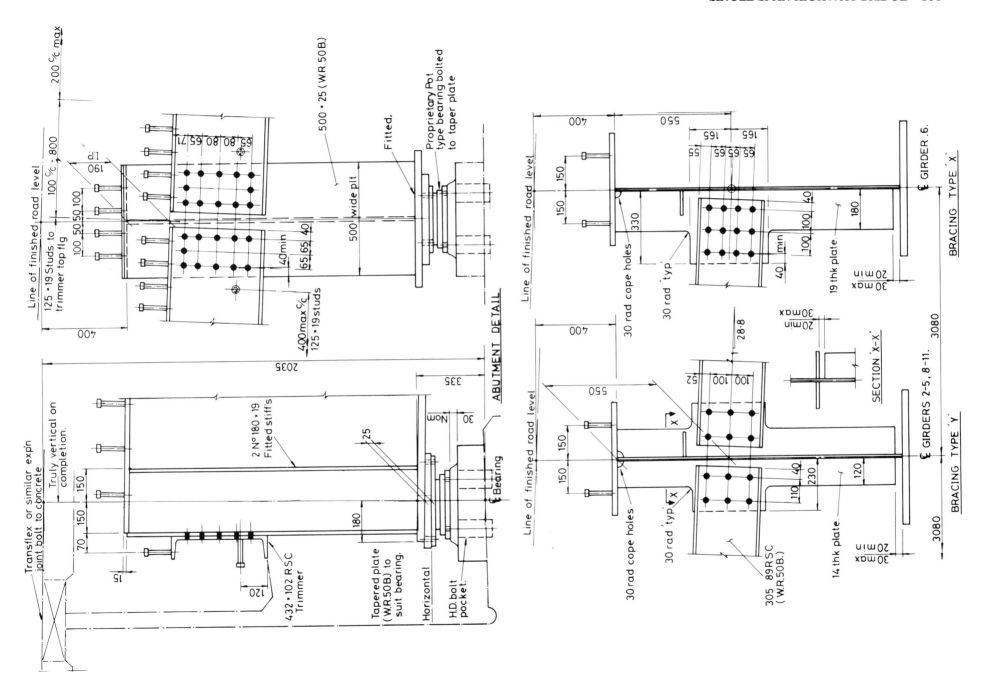

Figure 6.8.4 *Contd*

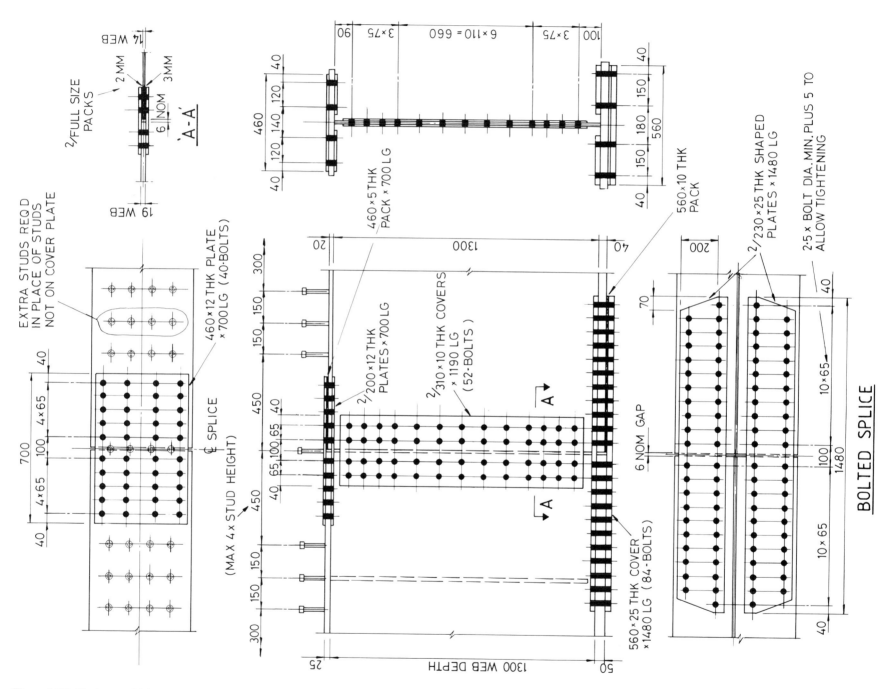

Figure 6.8.5 Single span highway bridge

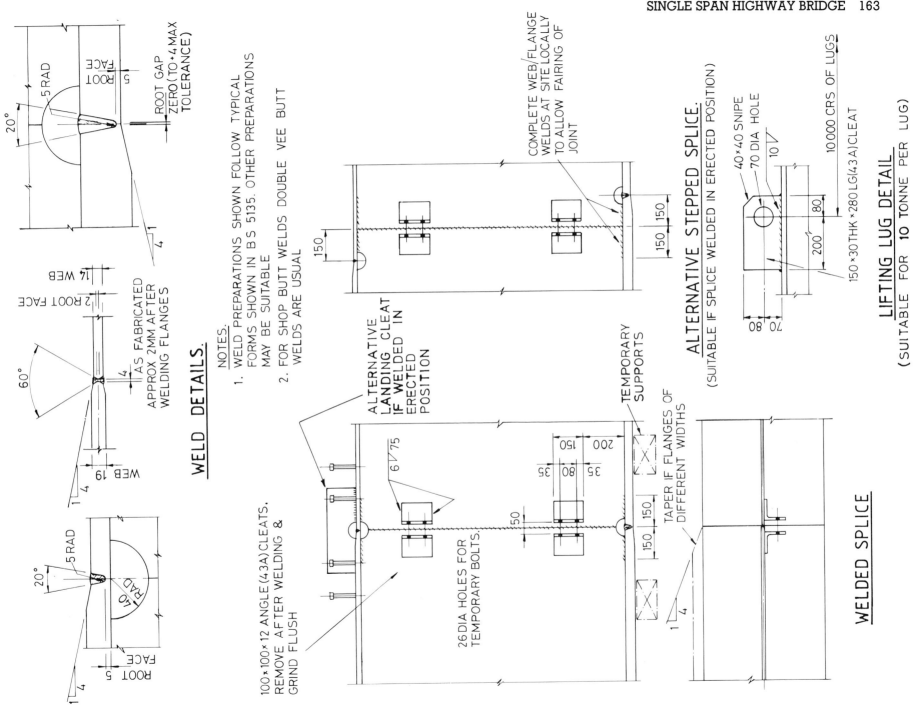

WELD DETAILS.

NOTES.
1. WELD PREPARATIONS SHOWN FOLLOW TYPICAL FORMS SHOWN IN B.S 5135. OTHER PREPARATIONS MAY BE SUITABLE
2. FOR SHOP BUTT WELDS DOUBLE VEE BUTT WELDS ARE USUAL

ALTERNATIVE STEPPED SPLICE.
(SUITABLE IF SPLICE WELDED IN ERECTED POSITION)

LIFTING LUG DETAIL
(SUITABLE FOR 10 TONNE PER LUG)

WELDED SPLICE

Figure 6.8.5 *Contd*

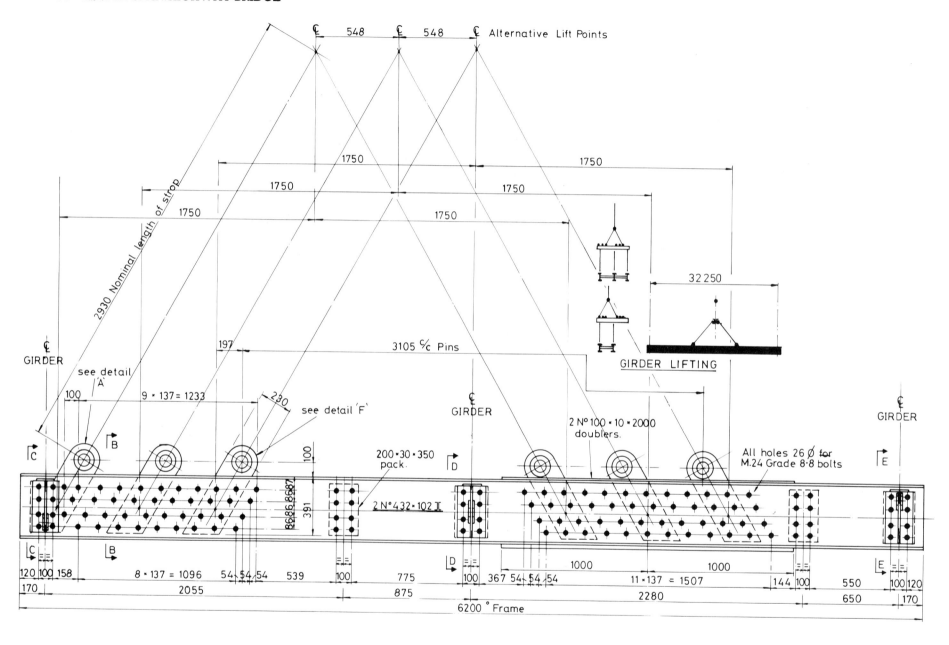

Figure 6.8.6 Single span highway bridge

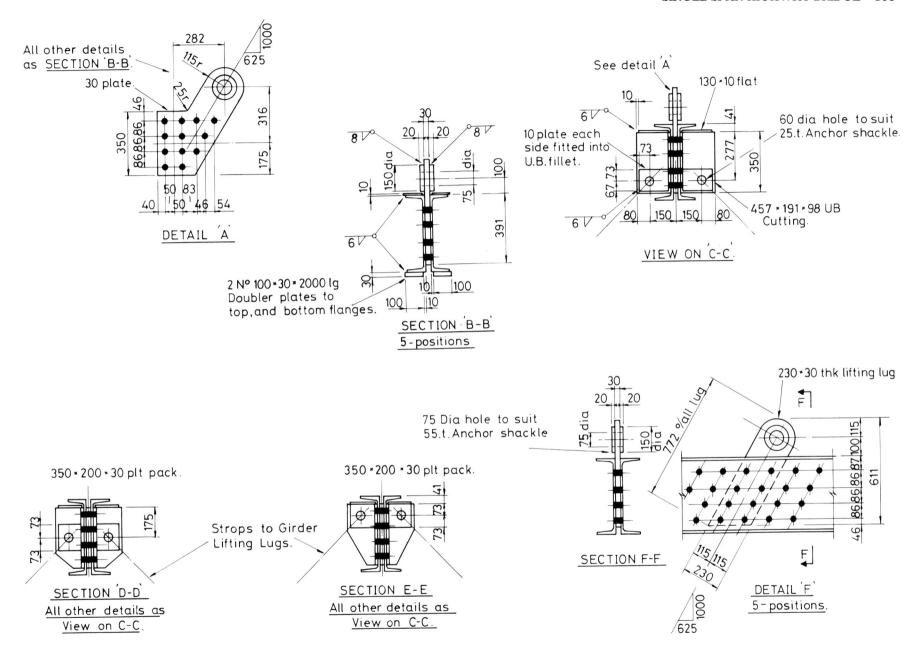

Figure 6.8.6 *Contd*

6.9 HIGHWAY SIGN GANTRY

The gantry displays destination signs (or traffic surveillance equipment) signals above a three-lane carriageway road. As shown it is suitable for mounting of internally illuminated signs. For larger directional signs as used on motorways external illumination is more usual with lighting units mounted on a walkway located in front of and below the signs. Such walkway could also be used for maintenance access and a heavier type of gantry results. Location is adjacent to the coast.

Rectangular hollow sections are used throughout to give a clean appearance. The legs suport a U-girder of Vierendeel form with the signs mounted within the rectangular openings bounded by suitably positioned vertical chords. A proprietary cable tray is carried which also serves as an access walkway. Sign cables are conveyed within the legs inside steel conduits so as to prevent damage or being unsightly. Welded joints are used throughout except for the leg to girder connections which are site bolted using tensioned screwed rods to ensure rigid portal action of the gantry.

The holding down bolt arrangement is designed to allow rapid erection during a night road closure. This is achieved using a 'bolt box' arrangement with loose top washer plates for tolerance. 'Finger' packs are supplied so that accurate levelling and securing of the gantry can be achieved, with final grouting of the bases later.

Drawing notes

(1) All steel to be to BS 4360:1986 grade 43A uos.
     Hollow sections to be grade 43C.
(2) Protective treatment – marine environment.
     Grit blast 1st quality after fabrication.
     Metal coating – aluminium spray
     Paint coats: 1st aluminium epoxy sealer
                  2nd zinc phosphate CR/alkyd undercoat
                  3rd zinc phospate CR/alkyd undercoat
                  4th MIO CR undercoat
                  5th CR finish.
     Minimum total dry film thickness 250 microns.
     [Department of Transport system type 9]

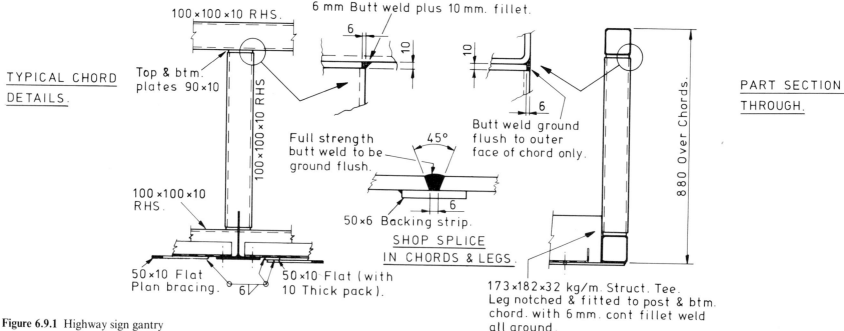

Figure 6.9.1 Highway sign gantry

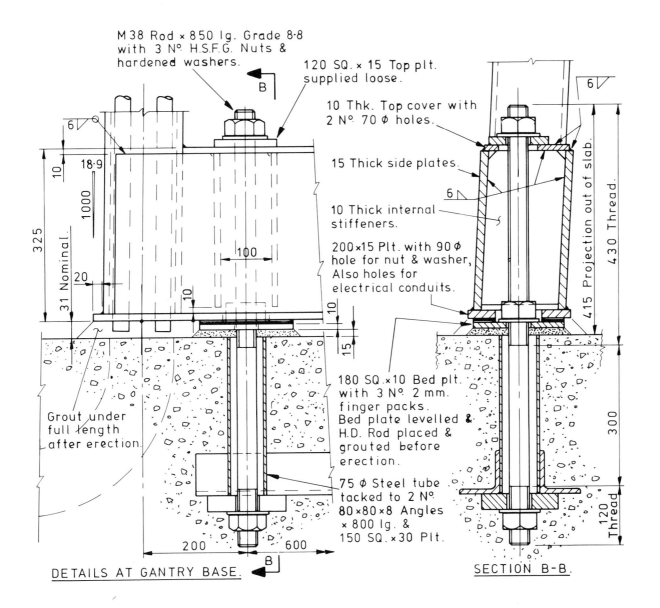

M 38 Rod × 850 lg. Grade 8·8 with 3 N°. H.S.F.G. Nuts & hardened washers.

120 SQ. × 15 Top plt. supplied loose.

10 Thk. Top cover with 2 N°. 70 Ø holes.

15 Thick side plates.

10 Thick internal stiffeners.

200×15 Plt. with 90 Ø hole for nut & washer, Also holes for electrical conduits.

180 SQ. ×10 Bed plt. with 3 N°. 2 mm. finger packs. Bed plate levelled & H.D. Rod placed & grouted before erection.

75 Ø Steel tube tacked to 2 N°. 80×80×8 Angles × 800 lg. & 150 SQ. ×30 Plt.

Grout under full length after erection.

325

31 Nominal.

1000

18·9

10

20

100

10

10

15

200

600

DETAILS AT GANTRY BASE.

6 Projection out of slab.

4·30 Thread.

415 Projection out of slab.

300

120 Thread.

SECTION B-B.

Figure 6.9.1 *Contd*

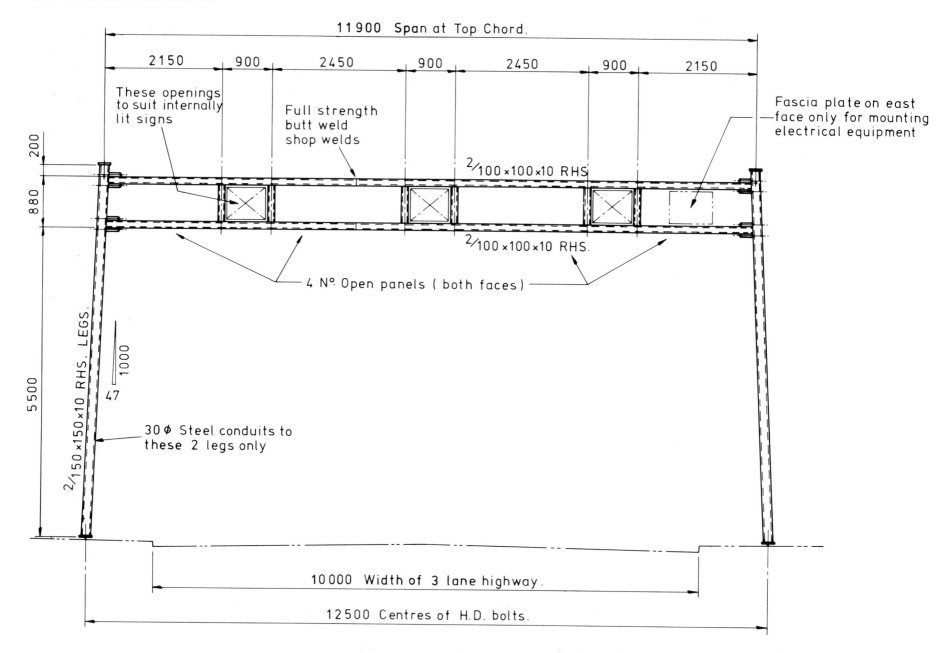

Figure 6.9.2 Highway sign gantry

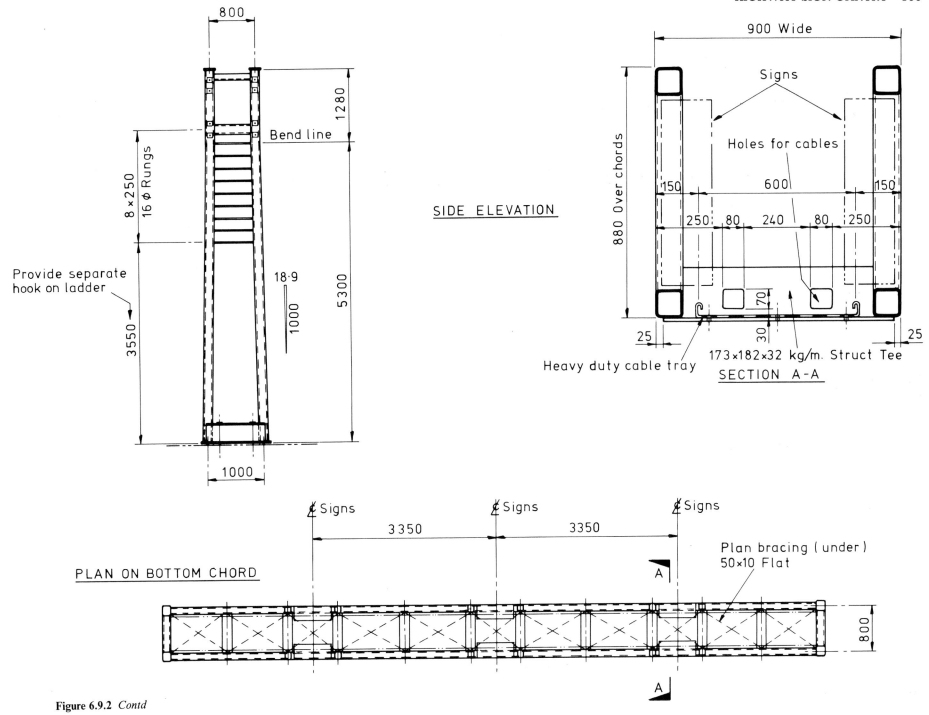

SIDE ELEVATION

800

1280

8 × 250
16 ⌀ Rungs

Bend line

18·9

1000

5300

3550

Provide separate
hook on ladder

1000

900 Wide

Signs

Holes for cables

880 Over chords

150   600   150

250 80 240 80 250

70

30

25   25

Heavy duty cable tray

173×182×32 kg/m. Struct Tee

SECTION A-A

₵ Signs   ₵ Signs   ₵ Signs

3350   3350

PLAN ON BOTTOM CHORD

Plan bracing (under)
50×10 Flat

A

A

800

Figure 6.9.2 *Contd*

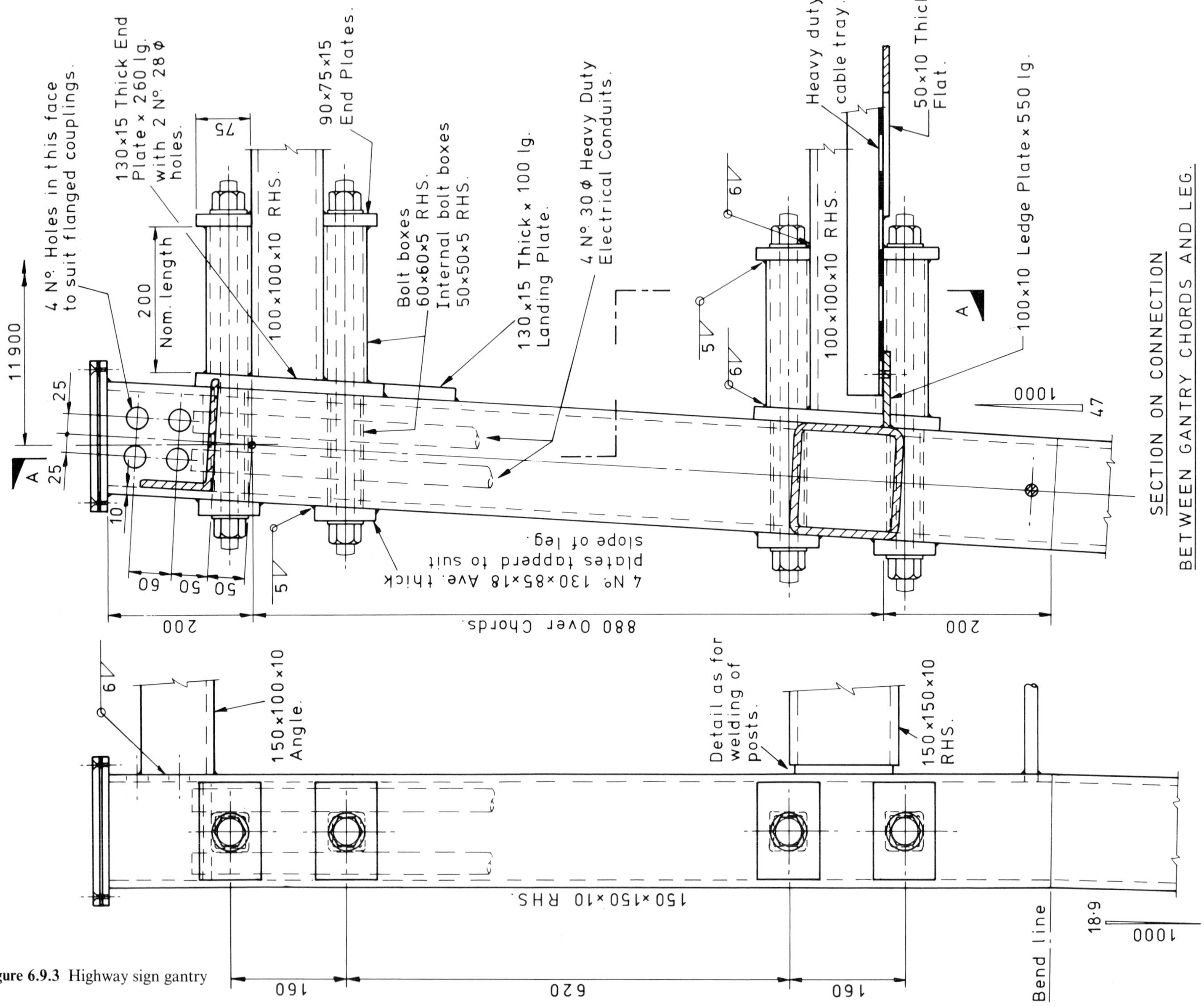

4 Nº. Holes in this face to suit flanged couplings.

130×15 Thick End Plate × 260 lg. with 2 Nº 28 ⌀ holes.

90×75×15 End Plates.

Bolt boxes 60×60×5 RHS. Internal bolt boxes 50×50×5 RHS.

130×15 Thick × 100 lg. Landing Plate.

4 Nº 30 ⌀ Heavy Duty Electrical Conduits.

Heavy duty cable tray.

50×10 Thick Flat.

100×10 Ledge Plate×550 lg.

75

200
Nom. length

100×100×10 RHS.

100×100×10 RHS.

11 900

25

25

10

60 50 50

5

4 Nº. 130×85×18 Ave thick plates tapperd to suit slope of leg.

880 Over Chords.

200

200

6

5

6

A

A

1000

47

SECTION ON CONNECTION

BETWEEN GANTRY CHORDS AND LEG.

6

150×100×10 Angle.

Detail as for welding of posts.

150×150×10 RHS.

150×150×10 RHS.

Bend line

160

620

160

18.9

1000

END VIEW.

Figure 6.9.3 Highway sign gantry

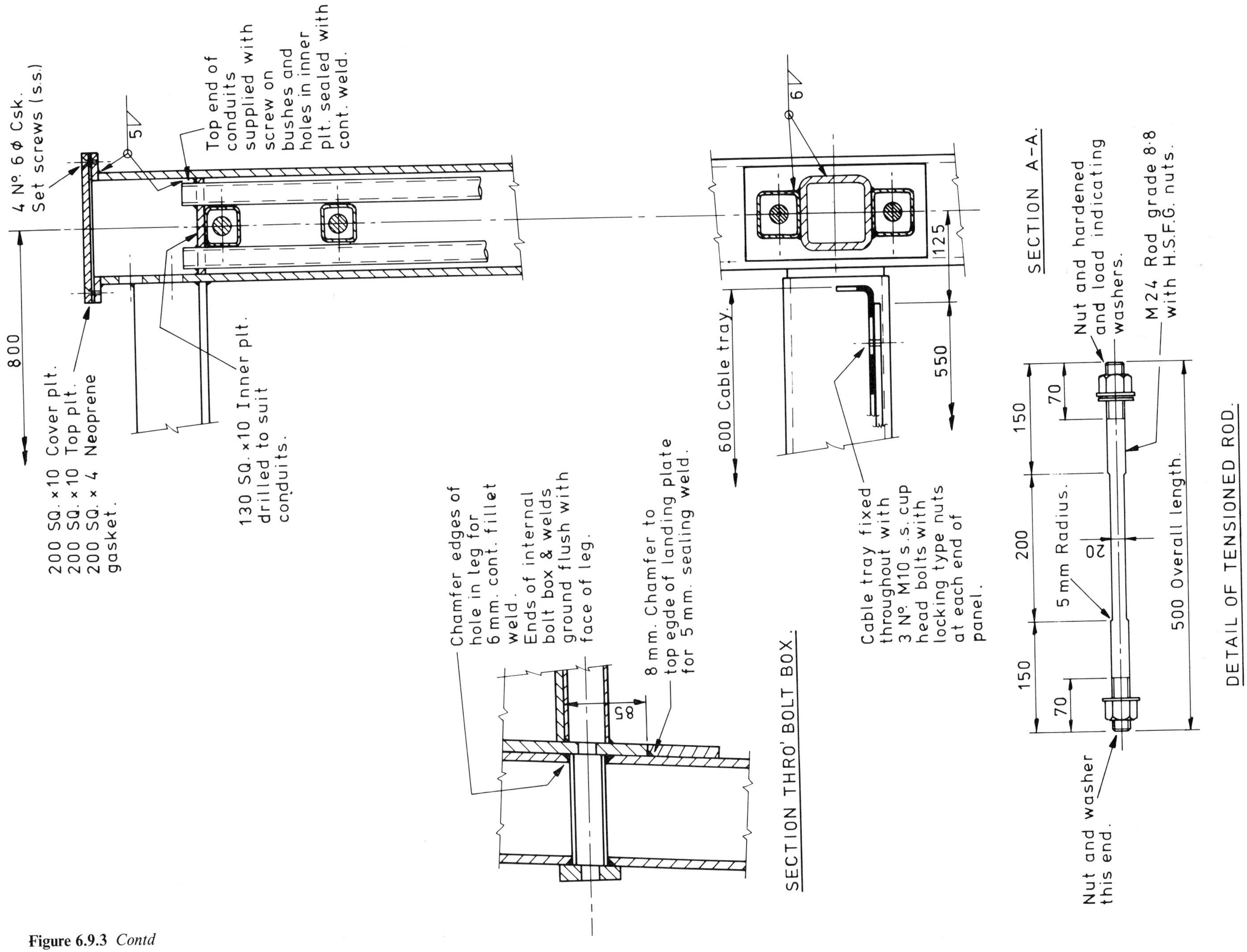

4 N° 6 φ Csk. Set screws (s.s.)

Top end of conduits supplied with screw on bushes and holes in inner plt. sealed with cont. weld.

200 SQ. ×10 Cover plt.
200 SQ. ×10 Top plt.
200 SQ. × 4 Neoprene gasket.

130 SQ. ×10 Inner plt. drilled to suit conduits.

800

600 Cable tray.

SECTION A-A.

550

Cable tray fixed throughout with 3 N° M10 s.s. cup head bolts with locking type nuts at each end of panel.

Chamfer edges of hole in leg for 6 mm. cont. fillet weld. Ends of internal bolt box & welds ground flush with face of leg.

8 mm. Chamfer to top egde of landing plate for 5 mm. sealing weld.

SECTION THRO' BOLT BOX.

Nut and hardened and load indicating washers.

M 24 Rod grade 8·8 with H.S.F.G. nuts.

150
70
200
5 mm Radius.
150
70
500 Overall length.

Nut and washer this end.

DETAIL OF TENSIONED ROD.

Figure 6.9.3 *Contd*

6.10 STAIRCASE

The staircase occurs within an industrial complex and is an essential structure. It is typical of many staircases built within factories and would be suitable as fire escape stairs in public buildings. Figure 6.10.1 shows one landing/flight unit which is connected to similar elements to form a zig-zag staircase. Certain design standards relate to staircases regarding proportions of rise: going, length of landings, number of risers between landings, etc. and these are shown in figure 4.1.

The staircase comprises twin steel flat stringers to which are bolted stair treads and tubular handrails. The stringers rely upon the treads to maintain stability against buckling. Channel stringers are also often used. Stair treads and floor/landing panels are of proprietary open bar grating type formed from a series of parallel flat load bearing bars stood on end and equispaced with either indented round or square bars. These are resistance welded into the top surface of the load bearing bars primarily to keep them upright. Panels typically 1 m wide and 6 m long or more are supported for elevated walkways and platforms. Normal treatment is galvanising which ensures that all intersces receive treatment, but between dip treatment can be used for less corrosive conditions. Stair treads are of similar construction. A number of manufacturers supply this type of flooring.

Handrail standards are proprietary solid forged type with tubular rails made from steel tube to BS 1775 grade 13. These are available from several manufacturers for either light or heavy duty applications.

GENERAL NOTES.
1. ALL STAIR STRINGER JOINTS TO BE FULL STRENGTH BUTT WELDS.
2. ALL OTHER WELDS TO BE 6 FILLET CONTINUOUS U.O.S.
3. ALL HOLES FOR HANDRAIL STANDARDS & STAIR TREADS TO BE 14 DIA. FOR M12 GRADE 4.6 BOLTS.
4. ALL OTHER HOLES ARE TO BE 22 DIA. FOR M20 GRADE 4.6 BOLTS.
5. ALL DIMENSIONS & DETAILS SHOWN FOR HANDRAILING, STANDARDS & TREADS ARE TO REDMAN FISHER STANDARD PATTERN.
6. STANDARDS :- 32 DIA. SOLID STEEL BARS.
7. HANDRAILING :- 25 DIA. NOM. BORE TUBES. (SPIGOT JOINTS TO BE ARRANGED AS REQD.)
8. STAIRTREADS :- REF. FD509 SERRATED LOAD BEARING BARS - 30×3 41 PITCH × 525 LONG, DIM.'A' = 249 & DIM.'B' = 100
9. FINISH - ALL MATERIALS TO BE GALVANISED EXCEPT FOR STAIR & LANDING STRINGERS, WHICH ARE PAINTED AS PER SPECIFICATION.
10. MATERIALS - TO BE TO BS 4360 (43A) U.O.S.

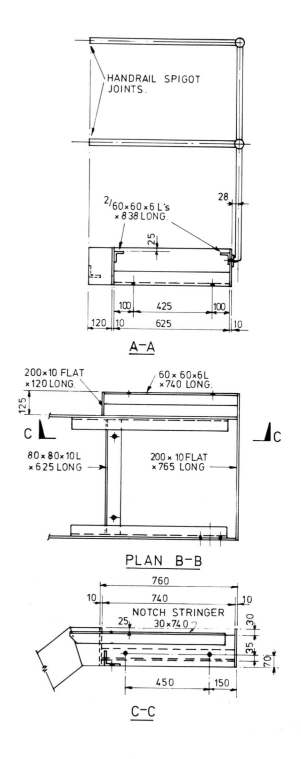

Figure 6.10.1 Staircase

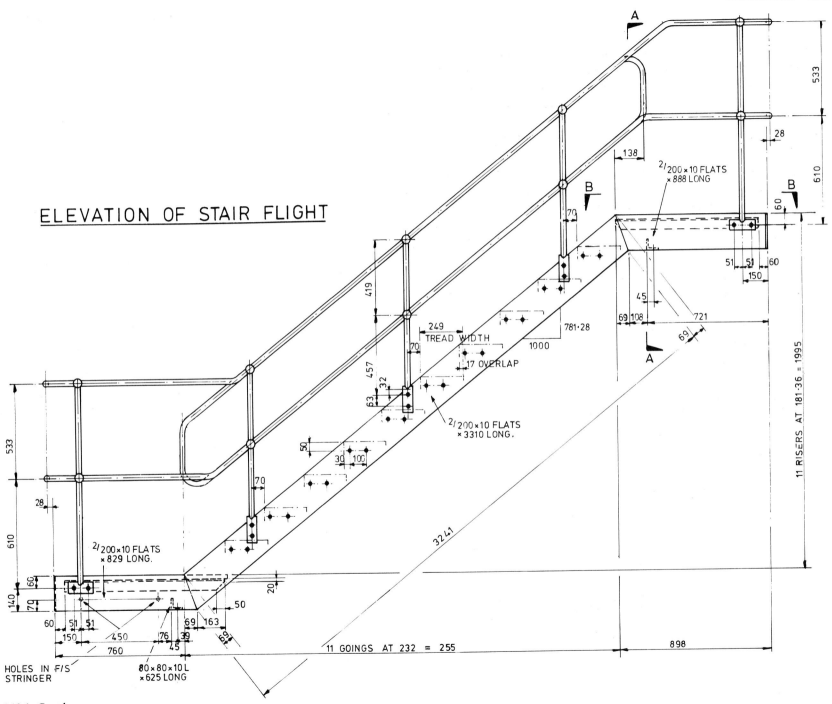

ELEVATION OF STAIR FLIGHT

Figure 6.10.1 *Contd*

SLIP RESISTANT FRONT
EDGE TO MATCH STAIR
TREAD REF. FD509

990 OVERALL PANEL WIDTH

41 PITCH

615 OVERALL PANEL LENGTH

102

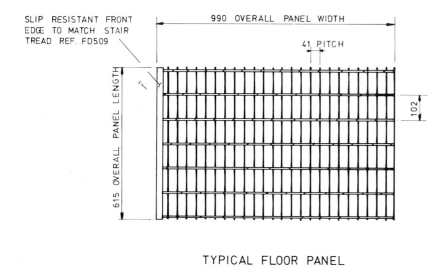

TYPICAL FLOOR PANEL

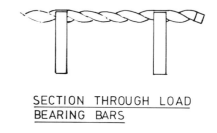

SECTION THROUGH LOAD
BEARING BARS

FLOOR PANEL SPECIFICATIONS

REDMAN FISHER :- FLOWFORGE OPEN STEEL FLOORING
TYPE 41/102 SERRATED LOAD BEARING BARS 25 x 5
4 FIXING CLIPS REQD. PER PANEL CLIP REF. FD500
(ALL MATERIALS GALVANISED)

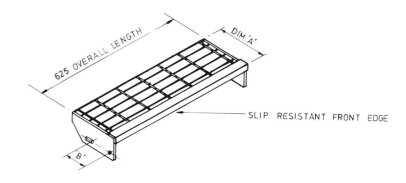

625 OVERALL LENGTH

DIM. 'A'

SLIP RESISTANT FRONT EDGE

'B'

STAIR TREAD DETAIL
REF. FD509

Figure 6.10.2 Staircase

REFERENCES

1 BS 4360:1986 *Weldable structural steels.*
2 BS 5950 *Structural use of steelwork in building.*
   Part 1:1985 Code of Practice for design in simple and continuous construction in hot rolled sections.
   Part 2:1985 Specification for materials, fabrication and erection:hot rolled sections.
   Part 4:1982 Code of Practice for design of floors with profiled steel sheeting.
3 BS 5400 *Steel, concrete and composite bridges.*
   Part 1:1978 General statement.
   Part 2:1978 Specification for loads.
   Part 3:1982 Code of Practice for design of steel bridges.
   Part 4:1984 Code of Practice for design of concrete bridges.
   Part 5:1979 Code of Practice for design of composite bridges.
   Part 6:1980 Specification for materials and workmanship, steel.
   Part 7:1978 Specification for materials and workmanship, concrete, reinforcement and prestressing tendons.
   Part 8: 1978 Recommendations for materials and workmanship, concrete, reinforcement and prestressing tendons.
   Part 9:1980 Bridge bearings.
   Part 10:1980 Code of Practice for fatigue.
4 Departmental Standard BD7/81. *Weathering steel for highway structures.* Department of Transport, 1981.
5 BS 4:Part1:1980 *Specification for hot rolled sections.*
6 BS 4848 *Hot rolled structural steel sections.*
   Part 2:1975 Hollow sections.
   Part 4:1972 Equal and unequal angles.
7 BS 5135:1984 *Specification for metal-arc welding of carbon and carbon-manganese steels.*
8 BS 4190:1967 *ISO metric black hexagon bolts, screws and nuts.*
9 BS 4320:1968 *Metal washers for general engineering purposes. Metric series.*
10 BS 3692:1967 *ISO metric precision hexagon bolts, screws and nuts.*
11 BS 4395 *High strength friction grip bolts and associated nuts and washers for structural engineering.*
    Part 1:1969 General grade.
    Part 2:1969 Higher grade.
    Part 3:1973 Higher grade (waisted shank).
12 BS 4604 *The use of high strength friction grip bolts in structural steelwork.*
    Part 1:1970 General grade.
    Part 2:1970 Higher grade.
    Part 3:1973 Higher grade (waisted shank).
13 BS 4232:1967 *Surface finish of blast-cleaned steel for painting.*
14 Swedish Standard SIS 05 59 00 *Rust grades for steel surfaces and preparation grades prior to protective coating.* Swedish Standards Commission, Stockholm, 1971.
15 *Steel structures painting manual.* Steel Structures Painting Council, Pittsburgh, USA.
    Volume 1:1966 Good painting practice.
    Volume 2:1973 Systems and specification.
16 BS 5493:1977 *Code of Practice for Protective coating of iron and steel structures against corrosion.*
17 *Notes of guidance on the specification for highway works* Part 6, Department of Transport, HMSO, 1986.
18 *Steelwork design guide to BS 5950:Part 1:1985.*

Volume 1: Section properties. Member capacities, Constrado (now SCI), 1985.

19 CP 117 *Composite construction in structural steel and concrete.*
Part 1:1965 Simply supported beams in buildings.

20 Steel framed multi-storey buildings. *Design recommendations for composite floors using steel decks.* Section 1:Structural, Constrado (now SCI). 1983.

21 BS 3429:1961 *Specification for sizes of drawing sheets.*

22 BS 639:1976 *Covered electrodes for the manual metal arc welding of carbon manganese steels.*

23 *Structural fasteners and their application,* BCSA, 1978.

24 Fire protection for structural steel in buildings, Constrado (now SCI), 1983.

FURTHER READING

DESIGN

(1) *Steel Designer's Manual* 4th Edn (revised), Constrado (now SCI), 1983.

(2) BS 6399 *Design loading for buildings*, Part 1:1984. *Code of Practice for dead and imposed loads* (formerly CP3:Chapter V:Part 1).

(3) BS 5502 *Code of Practice for the design of buildings and structures for agriculture*.
Part 1:Section 1.1:1978 Materials.
Part 1:Section 1.2:1980 Design construction and loading.

(4) BS 2573 *Rules of the design of cranes*.
Part 1:1983 Specification for classification of stress calculations and design criteria for structures.

(5) BS 2853:1959 *The design and testing of steel overhead runway beams*.

(6) BS 3579:1963 *Heavy duty electric overhead travelling and special cranes*.

(7) *Manual on connections for beam and column construction*, Publication No. 9/82, John W. Pask, BCSA, 1982.

(8) *Code of Practice for factory steel stairways, ladders and handrails*, Handbook 7 (revised 1973), Engineering Equipment Users Association.

(9) *Manual of steel construction*, 8th Edn, AISC, 1980.

(10) BS 5395:1977 *Code of Practice for stairs*.

(11) *Plastic design*, L.J. Morris & A.L. Randall, Constrado (now SCI), 1975.

DETAILING

(1) *Metric practice for structural steelwork*, Publication No. 5/79, 3rd Edn, BCSA, 1979.

(2) *Prefabricated floors for steel framed buildings*, BCSA, 1977.

(3) BS 308 *Engineering drawing practice*.
Part 1 General principles.
Part 2 Dimensions.
Part 3 Geometrical tolerancing.

(4) *Structural steel detailing*, AISE, 2nd Edn, 1971.

STEEL SECTIONS

(1) *Structural steelwork handbook*. Properties and safe load tables. BCSA/Constrado (now SCI), 1978.

(2) BS 1387:1967 *Steel tubes and tubulars*.

(3) *Piling handbook*. BSC, 1st Edn, 1976.

PROTECTIVE TREATMENT

(1) CP 1021 *Cathodic protection*. British Standards Institution.

(2) BS 729:1971 *Hot dip galvanised coatings on iron and steel articles*.

(3) BS 1706 *Electroplated coatings of cadmium and zinc on iron and steel*.

(4) BS 2569 *Sprayed metal coatings*.
Part 1:1964 Protection of iron and steel by aluminium and zinc against atmospheric corrosion.

(5) BS 3382:1961 *Electroplated coatings on threaded components*.
Part 1 Cadmium on steel components.
Part 2 Zinc on steel components.

(6) *Durability of steel structures*, Handbook by West European Steel Information Centre, 1982, UK sponsor Constrado (now SCI).

(7) *Steelwork corrosion protection guide – exterior environments*, Leaflet Ref. No. BSC S812 10.9.86, British Steel Corporation.

(8) *Corrosion in civil engineering*, Proceedings of the conference held in London, 1979, Institution of Civil Engineers.

ERECTION

(1) BS 5531:1978 *Safety in erecting structural frames.*

(2) *Safety in steel erection*, Publication ISBN 011 833 2417, HMSO, 1979.

(3) BS 5975:1982 *Code of Practice for falsework.*

COMPOSITE CONSTRUCTION

(1) *Composite structures of steel and concrete*, R.P. Johnson and R.J. Buckby.
Volume 1 Beams, columns, frames and applications in Building, 1979.
Volume 2 Bridges, 2nd Edn, 1986, Collins (now BSP).

BRIDGES

(1) *Steel bridges*, Derek Tordoff, BCSA publication No. 15/85.

(2) *International symposium on steel bridges*, ECCS/BCSA, Publication No. 53, London, 1988.

INTERNATIONAL

(1) *International structural steelwork handbook*, Publication No. 6/83, BCSA.

(2) *Iron and steel specifications*, 6th Edn, British Steel Corporation, 1986.

WELDING

(1) *Design of welded structures*, O.W. Blodgett and James F. Lincoln, Arc Welding Foundation, USA, 1966.

(2) *Introduction to the welding of structural steelwork*, J.L. Pratt, Constrado, 1979 (now SCI).

(3) ANSI/AWS D1, 1–81 *Structural welding code*, USA.

(4) BS 4165:1984 *Specification for electrode wires and fluxes for the submerged arc welding of carbon steel and medium tensile steels.*

(5) BS 449 *Welding terms and symbols.*
Part 1 Glossary of terms.
Part 2:1980 Symbols for welding.
Part 3 Terminology of weld imperfections.

WELD TESTING

(1) BS 4870 *Approval testing of welding procedures.*
Part 1:1981 Fusion welding of steel.

(2) BS 4871 *Approval testing of welders working to approved welding procedures.*
Part 1:1982 Fusion welding of steel.

(3) BS 4872 *Approval testing of welders when welding procedure approval is not required.*
Part 1:1982 Fusion welding of steel.

(4) BS 709:1983 *Methods of destructive testing of fusion welded joints and weld metals in steel.*

(5) BS 2600 *Radiographic examination of fusion welded butt joints in steel.*
Part 1:1983 Methods for steel 2 mm up to and including 50 mm thick.
Part 2:1973 Over 50 mm up to and including 200 mm thick.

(6) BS 3923 *Methods for ultrasonic examination of welds.*
Part 1:1978 Manual examination of fusion welds in ferric steels.

(7) BS 6072:1981 *Method for magnetic particle flaw detection.*

(8) BS 6443:1984 *Methods for penetrant flaw detection.*

(9) BS 5289 *Code of practice for the visual inspection of fusion welded joints.*

(10) PD 6493:1980 *Guidance on some methods for the derivation of acceptance levels for defects in fusion welded joints*, British Standards Institution.

ABBREVIATIONS

BS British Standard – British Standards may be obtained from: British Standards Institution, Sales Department, Linford Wood, Milton Keynes, MK14 6LE.

BCSA British Construction Steelwork Association Limited, 35 Old Queen Street, London, SW1.

SCI Steel Construction Institute, Silwood Park, Ascot, Berkshire, SL5 7PY.

AISC American Institute of Steel Construction, 101 Park Avenue, New York 17, New York, U.S.A.

HMSO Her Majesty's Stationery Office, HMSO Publication Centre, P.O. Box 276, London, SW8 5DT.

BSC British Steel Corporation,
Sections Structural Advisory Services, BSC Sections & Commercial Steels, P.O. Box 24, Steel House, Redcar, Cleveland, TS10 5QL.
Piling Piling Sections & Commercial Steels, Frodingham House, P.O. Box 1, Scunthorpe, South Humberside, DN16 1BP.
Plates Technical Advisory Service, BSC Plates, P.O. Box 30, Motherwell, Lanarkshire, ML1 1AA.
Tubes Divisional Sales Office, Corby, Northamptonshire, NN17 1UA.

APPENDIX

The Appendix contains useful information including weights of bars and flats, conversion factors and trigonometrical expressions.

MASS OF ROUND AND SQUARE BARS

Kilogrammes per linear metre

Dia. or side	Round	Square	Dia. or side	Round	Square	Dia. or side	Round	Square
mm	●	■	mm	●	■	mm	●	■
10	0.62	0.79	45	12.48	15.90	100	61.65	78.50
11	0.75	0.95	46	13.05	16.61	105	67.97	86.55
12	0.89	1.13	47	13.62	17.34	110	74.60	94.90
13	1.04	1.33	48	14.21	18.09	115	81.54	103.82
14	1.21	1.54	49	14.80	18.85	120	88.78	113.04
15	1.39	1.77	50	15.41	19.63	125	96.33	122.66
16	1.58	2.01	51	16.04	20.42	130	104.19	132.67
17	1.78	2.27	52	16.67	21.23	135	112.36	143.07
18	2.00	2.54	53	17.32	22.05	140	120.84	153.86
19	2.23	2.83	54	17.98	22.89	145	129.63	165.05
20	2.47	3.14	55	18.65	23.75	150	138.72	176.63
21	2.72	3.46	56	19.33	24.62	155	148.12	188.60
22	2.98	3.80	57	20.03	25.50	160	157.83	200.96
23	3.26	4.15	58	20.74	26.41	165	167.85	213.72
24	3.55	4.52	59	21.46	27.33	170	178.18	226.87
25	3.85	4.91	60	22.20	28.26	175	188.81	240.41
26	4.17	5.31	61	22.94	29.21	180	199.76	254.34
27	4.49	5.72	62	23.70	30.18	185	211.01	268.67
28	4.83	6.15	63	24.47	31.16	190	222.57	283.39
29	5.19	6.60	64	25.25	32.15	195	234.44	298.50
30	5.55	7.07	65	26.05	33.17	200	246.62	314.00
31	5.92	7.54	66	26.86	34.19	205	259.10	329.90
32	6.31	8.04	67	27.68	35.24	210	271.89	346.19
33	6.71	8.55	68	28.51	36.30	215	284.99	362.87
34	7.13	9.07	69	29.35	37.37	220	298.40	379.94
35	7.55	9.62	70	30.21	38.47	225	312.12	397.41
36	7.99	10.17	71	31.08	39.57	230	326.15	415.27
37	8.44	10.75	72	31.96	40.69	235	340.48	433.52
38	8.90	11.34	73	32.86	41.83	240	355.13	452.16
39	9.38	11.94	74	33.76	42.99	250	385.34	490.63
40	9.86	12.56	75	34.68	44.16	260	416.78	530.66
41	10.36	13.20	80	39.46	50.24	270	449.46	572.27
42	10.88	13.85	85	44.54	56.72	280	483.37	615.44
43	11.40	14.51	90	49.94	63.59	290	518.51	660.19
44	11.94	15.20	95	55.64	70.85	300	554.88	706.50

Suppliers should be consulted regarding availability of sizes.

MASS OF FLATS

Kilogrammes per linear metre

Width mm	\multicolumn{17}{c}{Thickness in millimetres}																
	1	2	3	4	5	6	7	8	9	10	15	20	25	30	40	50
5	0.04	0.08	0.12	0.16	0.20	0.24	0.27	0.31	0.35	0.39	0.59	0.79	0.98	1.18	1.57	1.96
10	0.08	0.16	0.24	0.31	0.39	0.47	0.55	0.63	0.71	0.79	1.18	1.57	1.96	2.36	3.14	3.93
15	0.12	0.24	0.35	0.47	0.59	0.71	0.82	0.94	1.06	1.18	1.77	2.36	2.94	3.53	4.71	5.89
20	0.16	0.31	0.47	0.63	0.79	0.94	1.10	1.26	1.41	1.57	2.36	3.14	3.93	4.71	6.28	7.85
25	0.20	0.39	0.59	0.79	0.98	1.18	1.37	1.57	1.77	1.96	2.94	3.93	4.91	5.89	7.85	9.81
30	0.24	0.47	0.71	0.94	1.18	1.41	1.65	1.88	2.12	2.36	3.53	4.71	5.89	7.07	9.42	11.8
35	0.27	0.55	0.82	1.10	1.37	1.65	1.92	2.20	2.47	2.75	4.12	5.50	6.87	8.24	11.0	13.7
40	0.31	0.63	0.94	1.26	1.57	1.88	2.20	2.51	2.83	3.14	4.71	6.28	7.85	9.42	12.6	15.7
45	0.35	0.71	1.06	1.41	1.77	2.12	2.47	2.83	3.18	3.53	5.30	7.07	8.83	10.6	14.1	17.7
50	0.39	0.79	1.18	1.57	1.96	2.36	2.75	3.14	3.53	3.93	5.89	7.85	9.81	11.8	15.7	19.6
55	0.43	0.86	1.30	1.73	2.16	2.59	3.02	3.45	3.89	4.32	6.48	8.64	10.8	13.0	17.3	21.6
60	0.47	0.94	1.41	1.88	2.36	2.83	3.30	3.77	4.24	4.71	7.07	9.42	11.8	14.1	18.8	23.6
65	0.51	1.02	1.53	2.04	2.55	3.06	3.57	4.08	4.59	5.10	7.65	10.2	12.8	15.3	20.4	25.5
70	0.55	1.10	1.65	2.20	2.75	3.30	3.85	4.40	4.95	5.50	8.24	11.0	13.7	16.5	22.0	27.5
75	0.59	1.18	1.77	2.36	2.94	3.53	4.12	4.71	5.30	5.89	8.83	11.8	14.7	17.7	23.6	29.4
80	0.63	1.26	1.88	2.51	3.14	3.77	4.40	5.02	5.65	6.28	9.42	12.6	15.7	18.8	25.1	31.4
85	0.67	1.33	2.00	2.67	3.34	4.00	4.67	5.34	6.01	6.67	10.0	13.3	16.7	20.0	26.7	33.4
90	0.71	1.41	2.12	2.83	3.53	4.24	4.95	5.65	6.36	7.07	10.6	14.1	17.7	21.2	28.3	35.3
95	0.75	1.49	2.24	2.98	3.73	4.47	5.22	5.97	6.71	7.46	11.2	14.9	18.6	22.4	29.8	37.3
100	0.79	1.57	2.36	3.14	3.93	4.71	5.50	6.28	7.07	7.85	11.8	15.7	19.6	23.6	31.4	39.3
110	0.86	1.73	2.59	3.45	4.32	5.18	6.04	6.91	7.77	8.64	13.0	17.3	21.6	25.9	34.5	43.2
120	0.94	1.88	2.83	3.77	4.71	5.65	6.59	7.54	8.48	9.42	14.1	18.8	23.6	28.3	37.7	47.1
130	1.02	2.04	3.06	4.08	5.10	6.12	7.14	8.16	9.18	10.2	15.3	20.4	25.5	30.6	40.8	51.0
140	1.10	2.20	3.30	4.40	5.50	6.59	7.69	8.79	9.89	11.0	16.5	22.0	27.5	33.0	44.0	55.0
150	1.18	2.36	3.53	4.71	5.89	7.07	8.24	9.42	10.6	11.8	17.7	23.6	29.4	35.3	47.1	58.9
160	1.26	2.51	3.77	5.02	6.28	7.54	8.79	10.0	11.3	12.6	18.8	25.1	31.4	37.7	50.2	62.8
170	1.33	2.67	4.00	5.34	6.67	8.01	9.34	10.7	12.0	13.3	20.0	26.7	33.4	40.0	53.4	66.7
180	1.41	2.83	4.24	5.65	7.07	8.48	9.89	11.3	12.7	14.1	21.2	28.3	35.3	42.4	56.5	70.7
190	1.49	2.98	4.47	5.97	7.46	8.95	10.4	11.9	13.4	14.9	22.4	29.8	37.3	44.7	59.7	74.6
200	1.57	3.14	4.71	6.28	7.85	9.42	11.0	12.6	14.1	15.7	23.6	31.4	39.3	47.1	62.8	78.5
210	1.65	3.30	4.95	6.59	8.24	9.89	11.5	13.2	14.8	16.5	24.7	33.0	41.2	49.5	65.9	82.4
220	1.73	3.45	5.18	6.91	8.64	10.4	12.1	13.8	15.5	17.3	25.9	34.5	43.2	51.8	69.1	86.4
230	1.81	3.61	5.42	7.22	9.03	10.8	12.6	14.4	16.2	18.1	27.1	36.1	45.1	54.2	72.2	90.3
240	1.88	3.77	5.65	7.54	9.42	11.3	13.2	15.1	17.0	18.8	28.3	37.7	47.1	56.5	75.4	94.2
250	1.96	3.93	5.89	7.85	9.81	11.8	13.7	15.7	17.7	19.6	29.4	39.3	49.1	58.9	78.5	98.1

For actual widths and thicknesses available, application should be made to manufacturers.

Masses for greater widths and/or thicknesses than those tabulated may be obtained by appropriate addition from the range of masses given.

MASS OF FLATS *Contd*

Width							Thickness in millimetres									
mm	1	2	3	4	5	6	7	8	9	10	15	20	25	30	40	50
260	2.04	4.08	6.12	8.16	10.2	12.2	14.3	16.3	18.4	20.4	30.6	40.8	51.0	61.2	81.6	102
270	2.12	4.24	6.36	8.48	10.6	12.7	14.8	17.0	19.1	21.2	31.8	42.4	53.0	63.6	84.8	106
280	2.20	4.40	6.59	8.79	11.0	13.2	15.4	17.6	19.8	22.0	33.0	44.0	55.0	65.9	87.9	110
290	2.28	4.55	6.83	9.11	11.4	13.7	15.9	18.2	20.5	22.8	34.1	45.5	56.9	68.3	91.1	114
300	2.36	4.71	7.07	9.42	11.8	14.1	16.5	18.8	21.2	23.6	35.3	47.1	58.9	70.7	94.2	118
310	2.43	4.87	7.30	9.73	12.2	14.6	17.0	19.5	21.9	24.3	36.5	48.7	60.8	73.0	97.3	122
320	2.51	5.02	7.54	10.0	12.6	15.1	17.6	20.1	22.6	25.1	37.7	50.2	62.8	75.4	100	126
330	2.59	5.18	7.77	10.4	13.0	15.5	18.1	20.7	23.3	25.9	38.9	51.8	64.8	77.7	104	130
340	2.67	5.34	8.01	10.7	13.3	16.0	18.7	21.4	24.0	26.7	40.0	53.4	66.7	80.1	107	133
350	2.75	5.50	8.24	11.0	13.7	16.5	19.2	22.0	24.7	27.5	41.2	55.0	68.7	82.4	110	137
360	2.83	5.65	8.48	11.3	14.1	17.0	19.8	22.6	25.4	28.3	42.4	56.5	70.7	84.8	113	141
370	2.90	5.81	8.71	11.6	14.5	17.4	20.3	23.2	26.1	29.0	43.6	58.1	72.6	87.1	116	145
380	2.98	5.97	8.95	11.9	14.9	17.9	20.9	23.9	26.8	29.8	44.7	59.7	74.6	89.5	119	149
390	3.06	6.12	9.18	12.2	15.3	18.4	21.4	24.5	27.6	30.6	45.9	61.2	76.5	91.8	122	153
400	3.14	6.28	9.42	12.6	15.7	18.8	22.0	25.1	28.3	31.4	47.1	62.8	78.5	94.2	126	157
410	3.22	6.44	9.66	12.9	16.1	19.3	22.5	25.7	29.0	32.2	48.3	64.4	80.5	96.6	129	161
420	3.30	6.59	9.89	13.2	16.5	19.8	23.1	26.4	29.7	33.0	49.5	65.9	82.4	98.9	132	165
430	3.38	6.75	10.1	13.5	16.9	20.3	23.6	27.0	30.4	33.8	50.6	67.5	84.4	101	135	169
440	3.45	6.91	10.4	13.8	17.3	20.7	24.2	27.6	31.1	34.5	51.8	69.1	86.4	104	138	173
450	3.53	7.07	10.6	14.1	17.7	21.2	24.7	28.3	31.8	35.3	53.0	70.7	88.3	106	141	177
460	3.61	7.22	10.8	14.4	18.1	21.7	25.3	28.9	32.5	36.1	54.2	72.2	90.3	108	144	181
470	3.69	7.38	11.1	14.8	18.4	22.1	25.8	29.5	33.2	36.9	55.3	73.8	92.2	111	148	184
480	3.77	7.54	11.3	15.1	18.8	22.6	26.4	30.1	33.9	37.7	56.5	75.4	94.2	113	151	188
490	3.85	7.69	11.5	15.4	19.2	23.1	26.9	30.8	34.6	38.5	57.7	76.9	96.2	115	154	192
500	3.93	7.85	11.8	15.7	19.6	23.6	27.5	31.4	35.3	39.3	58.9	78.5	98.1	118	157	196
510	4.00	8.01	12.0	16.0	20.0	24.0	28.0	32.0	36.0	40.0	60.1	80.1	100	120	160	200
520	4.08	8.16	12.2	16.3	20.4	24.5	28.6	32.7	36.7	40.8	61.2	81.6	102	122	163	204
530	4.16	8.32	12.5	16.6	20.8	25.0	29.1	33.3	37.4	41.6	62.4	83.2	104	125	166	208
540	4.24	8.48	12.7	17.0	21.2	25.4	29.7	33.9	38.2	42.4	63.6	84.8	106	127	170	212
550	4.32	8.64	13.0	17.3	21.6	25.9	30.2	34.5	38.9	43.2	64.8	86.4	108	130	173	216
560	4.40	8.79	13.2	17.6	22.0	26.4	30.8	35.2	39.6	44.0	65.9	87.9	110	132	176	220
570	4.47	8.95	13.4	17.9	22.4	26.8	31.3	35.8	40.3	44.7	67.1	89.5	112	134	179	224
580	4.55	9.11	13.7	18.2	22.8	27.3	31.9	36.4	41.0	45.5	68.3	91.1	114	137	182	228
590	4.63	9.26	13.9	18.5	23.2	27.8	32.4	37.1	41.7	46.3	69.5	92.6	116	139	185	232
600	4.71	9.42	14.1	18.8	23.6	28.3	33.0	37.7	42.4	47.1	70.7	94.2	118	141	188	236

MASS OF FLATS

Kilogrammes per linear metre

Width	Thickness in millimetres															
mm	1	2	3	4	5	6	7	8	9	10	15	20	25	30	40	50
610	4.79	9.58	14.4	19.2	23.9	28.7	33.5	38.3	43.1	47.9	71.8	95.8	120	144	192	239
620	4.87	9.73	14.6	19.5	24.3	29.2	34.1	38.9	43.8	48.7	73.0	97.3	122	146	195	243
630	4.95	9.89	14.8	19.8	24.7	29.7	34.6	39.6	44.5	49.5	74.2	98.9	124	148	198	247
640	5.02	10.0	15.1	20.1	25.1	30.1	35.2	40.2	45.2	50.2	75.4	100	126	151	201	251
650	5.10	10.2	15.3	20.4	25.5	30.6	35.7	40.8	45.9	51.0	76.5	102	128	153	204	255
660	5.18	10.4	15.5	20.7	25.9	31.1	36.3	41.4	46.6	51.8	77.7	104	130	155	207	259
670	5.26	10.5	15.8	21.0	26.3	31.6	36.8	42.1	47.3	52.6	78.9	105	131	158	210	263
680	5.34	10.7	16.0	21.4	26.7	32.0	37.4	42.7	48.0	53.4	80.1	107	133	160	214	267
690	5.42	10.8	16.2	21.7	27.1	32.5	37.9	43.3	48.7	54.2	81.2	108	135	162	217	271
700	5.50	11.0	16.5	22.0	27.5	33.0	38.5	44.0	49.5	55.0	82.4	110	137	165	220	275
710	5.57	11.1	16.7	22.3	27.9	33.4	39.0	44.6	50.2	55.7	83.6	111	139	167	223	279
720	5.65	11.3	17.0	22.6	28.3	33.9	39.6	45.2	50.9	56.5	84.8	113	141	170	226	283
730	5.73	11.5	17.2	22.9	28.7	34.4	40.1	45.8	51.6	57.3	86.0	115	143	172	229	287
740	5.81	11.6	17.4	23.2	29.0	34.9	40.7	46.5	52.3	58.1	87.1	116	145	174	232	290
750	5.89	11.8	17.7	23.6	29.4	35.3	41.2	47.1	53.0	58.9	88.3	118	147	177	236	294
760	5.97	11.9	17.9	23.9	29.8	35.8	41.8	47.7	53.7	59.7	89.5	119	149	179	239	298
770	6.04	12.1	18.1	24.2	30.2	36.3	42.3	48.4	54.4	60.4	90.7	121	151	181	242	302
780	6.12	12.2	18.4	24.5	30.6	36.7	42.9	49.0	55.1	61.2	91.8	122	153	184	245	306
790	6.20	12.4	18.6	24.8	31.0	37.2	43.4	49.6	55.8	62.0	93.0	124	155	186	248	310
800	6.28	12.6	18.8	25.1	31.4	37.7	44.0	50.2	56.5	62.8	94.2	126	157	188	251	314
810	6.36	12.7	19.1	25.4	31.8	38.2	44.5	50.9	57.2	63.6	95.4	127	159	191	254	318
820	6.44	12.9	19.3	25.7	32.2	38.6	45.1	51.5	57.9	64.4	96.6	129	161	193	257	322
830	6.52	13.0	19.5	26.1	32.6	39.1	45.6	52.1	58.6	65.2	97.7	130	163	195	261	326
840	6.59	13.2	19.8	26.4	33.0	39.6	46.2	52.8	59.3	65.9	98.9	132	165	198	264	330
850	6.67	13.3	20.0	26.7	33.4	40.0	46.7	53.4	60.1	66.7	100	133	167	200	267	334
860	6.75	13.5	20.3	27.0	33.8	40.5	47.3	54.0	60.8	67.5	101	135	169	203	270	338
870	6.83	13.7	20.5	27.3	34.1	41.0	47.8	54.6	61.5	68.3	102	137	171	205	273	341
880	6.91	13.8	20.7	27.6	34.5	41.4	48.4	55.3	62.2	69.1	104	138	173	207	276	345
890	6.99	14.0	21.0	27.9	34.9	41.9	48.9	55.9	62.9	69.9	105	140	175	210	279	349
900	7.07	14.1	21.2	28.3	35.3	42.4	49.5	56.5	63.6	70.7	106	141	177	212	283	353
910	7.14	14.3	21.4	28.6	35.7	42.9	50.0	57.1	64.3	71.4	107	143	179	214	286	357
920	7.22	14.4	21.7	28.9	36.1	43.3	50.6	57.8	65.0	72.2	108	144	181	217	289	361
930	7.30	14.6	21.9	29.2	36.5	43.8	51.1	58.4	65.7	73.0	110	146	183	219	292	365
940	7.38	14.8	22.1	29.5	36.9	44.3	51.7	59.0	66.4	73.8	111	148	184	221	295	369
950	7.46	14.9	22.4	29.8	37.3	44.7	52.2	59.7	67.1	74.6	112	149	186	224	298	373

For actual widths and thicknesses available, application should be made to manufacturers.
Masses for greater widths and/or thicknesses than those tabulated may be obtained by appropriate addition from the range of masses given.

MASS OF FLATS *Contd*

Kilogrammes per linear metre

Width	Thickness in millimetres															
mm	1	2	3	4	5	6	7	8	9	10	15	20	25	30	40	50
950	7.54	15.1	22.6	30.1	37.7	45.2	52.8	60.3	67.8	75.4	113	151	188	226	301	377
970	7.61	15.2	22.8	30.5	38.1	45.7	53.3	60.9	68.5	76.1	114	152	190	228	305	381
980	7.69	15.4	23.1	30.8	38.5	46.2	53.9	61.5	69.2	76.9	115	154	192	231	308	385
990	7.77	15.5	23.3	31.1	38.9	46.6	54.4	62.2	69.9	77.7	117	155	194	233	311	389
1000	7.85	15.7	23.6	31.4	39.3	47.1	55.0	62.8	70.7	78.5	118	157	196	236	314	393
1020	8.01	16.0	24.0	32.0	40.0	48.0	56.0	64.1	72.1	80.1	120	160	200	240	320	400
1040	8.16	16.3	24.5	32.7	40.8	49.0	57.1	65.3	73.5	81.6	122	163	204	245	327	408
1060	8.32	16.6	25.0	33.3	41.6	49.9	58.2	66.6	74.9	83.2	125	166	208	250	333	416
1080	8.48	17.0	25.4	33.9	42.4	50.9	59.3	67.8	76.3	84.8	127	170	212	254	339	424
1100	8.64	17.3	25.9	34.5	43.2	51.8	60.4	69.1	77.7	86.4	130	173	216	259	345	432
1120	8.79	17.6	16.4	35.2	44.0	52.8	61.5	70.3	79.1	87.9	132	176	220	264	352	440
1140	8.95	17.9	26.8	35.8	44.7	53.7	62.6	71.6	80.5	89.5	134	179	224	268	358	447
1160	9.11	18.2	27.3	36.4	45.5	54.6	63.7	72.8	82.0	91.1	137	182	228	273	364	455
1180	9.26	18.5	27.8	37.1	46.3	55.6	64.8	74.1	83.4	92.6	139	185	232	278	371	463
1200	9.42	18.8	28.3	37.7	47.1	56.5	65.9	75.4	84.8	94.2	141	188	236	283	377	471
1220	9.58	19.2	28.7	38.3	47.9	57.5	67.0	76.6	86.2	95.8	144	192	239	287	383	479
1240	9.73	19.5	29.2	38.9	48.7	58.4	68.1	77.9	87.6	97.3	146	195	243	292	389	487
1260	9.89	19.8	29.7	39.6	49.5	59.3	69.2	79.1	89.0	98.9	148	198	247	297	396	495
1280	10.0	20.1	30.1	40.2	50.2	60.3	70.3	80.4	90.4	100	151	201	251	301	402	502
1300	10.2	20.4	30.6	40.8	51.0	61.2	71.4	81.6	91.8	102	153	204	255	306	408	510
1320	10.4	20.7	31.1	41.4	51.8	62.2	72.5	82.9	93.3	104	155	207	259	311	414	518
1340	10.5	21.0	31.6	42.1	52.6	63.1	73.6	84.2	94.7	105	158	210	263	316	421	526
1360	10.7	21.4	32.0	42.7	53.4	64.1	74.7	85.4	96.1	107	160	214	267	320	427	534
1380	10.8	21.7	32.5	43.3	54.2	65.0	75.8	86.7	97.5	108	162	217	271	325	433	542
1400	11.0	22.0	33.0	44.0	55.0	65.9	76.9	87.9	98.9	110	165	220	275	330	440	550
1420	11.1	22.3	33.4	44.6	55.7	66.9	78.0	88.2	100	111	167	223	279	334	446	557
1440	11.3	22.6	33.9	45.2	56.5	67.8	79.1	90.4	102	113	170	226	283	339	452	565
1460	11.5	22.9	34.4	45.8	57.3	68.8	80.2	91.7	103	115	172	229	287	344	458	573
1480	11.6	23.2	34.9	46.5	58.1	69.7	81.3	92.9	105	116	174	232	290	349	465	581
1500	11.8	23.6	35.3	47.1	58.9	70.7	82.4	94.2	106	118	177	236	294	353	471	589
1600	12.6	25.1	37.7	50.2	62.8	75.4	87.9	100	113	126	188	251	314	377	502	628
1700	13.3	26.7	40.0	53.4	66.7	80.1	93.4	107	120	133	200	267	334	400	534	667
1800	14.1	28.3	42.4	56.6	70.7	84.8	98.9	113	127	141	212	283	353	424	565	707
1900	14.9	29.8	44.7	59.7	74.6	89.5	104	119	134	149	224	298	373	447	597	746
2000	15.7	31.4	47.1	62.8	78.5	94.2	110	126	141	157	236	314	393	471	628	785

METRIC CONVERSION OF UNITS

Measure	From metric			To metric		
	Unit	Conversion		Unit	Conversion	
Length	mm	0.03937 in		in	25.4 mm (exact)	2.54 cm
	m	3.281 ft	0.5468 fathom	ft	0.3048 m	304.8 mm
		1.094 yd		yd	0.9144 m (exact)	
	km	0.6214 mile		mile	1.609 km	
Thickness	micron (μm)	0.03937 thou	(mil or milli-inch)	thou (milli-inch)	25.4 micron or micrometre	0.0254 mm
Area	mm^2	0.00155 m^2		m^2	645.2 m^2	6.452 cm^2
	cm^2	0.1550 in^2				
	m^2	10.76 ft^2	1.196 yd^{22}	ft^2	0.0929 m^2	
	hectare (100 m $\times$ 100 m)	2.471 acres		acre	0.4047 hectares	
	km^2	0.3861 sq. miles		sq. miles	2.590 km^2	
Volume	mm^3	0.00006102 m^3		in^3	16,390 mm^3	16.39 cc
	m^3	35.31 ft^3	1.308 yd^3	ft^3	0.02832 m^3	
Capacity	litre	0.22 imp. gallons	0.2542 US gallons	gallon	4.546 litre	(10 lb of water)
				US gallon	3.785 litre	
		(1 US gallons = 0.8327 imp gall)		pint	0.568 litre	
Mass	tonne (1000 kg)	0.9842 ton	1.102 USA short ton (2000 lb)	ton (2240 lb)	1016 kg	1.016 tonne
	kg (1000 kg)	2.205 lb		lb	0.4536 kg	
	g	0.03527 oz		oz	28.35 g	
Density	kg/m^3	0.0624 lb/ft^3		lb/ft^3	16.02 kg/m^3	
Force	N (Newton)	0.2248 lbf	0.1020 kgf	lbf	4.448 N	0.4536 kgf
	kgf	2.205 lbf	9.807 N			
	kN	0.1004 tonf	0.1020 tonnef	tonf	9.964 kN	1.016 tonnef
		0.2248 Kip (US)				
	tonnef (1000 kgf)	0.9842 tonf	9.807 kN	Kip (US) (1000 lbf)	4.448 kN	

METRIC CONVERSION OF UNITS *Contd*

Measure	From metric		To metric	
	Unit	Conversion	Unit	Conversion
Force per unit length	N/m kN/m (or N/mm) tonnef/m	0.06852 lbf/ft 0.1020 kgf/m 0.0306 tonf/ft 0.1020 tonnef/m 0.00255 tonf/m 9.807 kN/m 0.3000 tonf/ft	lbf/ft tonf/ft tonf/in	14.59 N/m 1.488 kgf/m 32.69 kN/m 3.333 tonnef/m 392 kN/m
Pressure stress or modulus of elasticity	kN/m^2 kg/cm^2 N/mm^2	0.009324 $tonf/ft^2$ 0.01020 kgf/m^2 0.9144 $tonf/ft^2$ 98.07 kN/m^2 145.0 lbf/in^2 10.20 kgf/cm^2 0.145 ksi 0.06475 $tonf/m^2$ 10 bar 1000 000 pascal	$tonf/ft^2$ lbf/in^2 (psi) $tonf/in^2$	107.3 kN/m^2 1.094 kg/cm^3 0.006895 N/mm^2 0.0703 kgf/cm^2 15.44 N/m^2 157.5 kgf/cm^2
	kgf/cm^2	14.22 lbf/m^2 0.09807 N/m^2 0.006350 $tonf/m^2$		
	atm (standard atmosphere)	14.70 lbf/in^2	lb/in^2	0.06805 atm
Moment	kN/m N/m kgf/m	0.3293 tonf/ft 0.7376 lbf/ft 0.1020 kgf/m 7.233 lbg/ft 9.807 N/m	tonf/ft lbf/ft	3.037 kN/m 1.356 N/m 0.1283 kgf/m
Section Modulus (z)	cm^3	0.06102 m^3	in^3	16.39 cm^3
Second moment of area (I)	cm^4	0.02403 m^4	in^4	41.62 cm^4
Acceleration	m/sec^2	3.281 ft/sec^2	ft/sec^2	0.3048 m/sec^2
Gravity acceleration	9.807 m/sec^2		32.17 ft/sec^2	
Velocity	km/hr m/sec	0.6214 mph 0.5396 UK Knots 3.281 ft/sec	mph ft/sec UK Knot	1.609 km/hr 0.3048 m/sec 1.853 km/hr

METRIC CONVERSION OF UNITS *Contd*

Measure	From metric		To metric	
	Unit	Conversion	Unit	Conversion
Temperature	°C	$(°F - 32) \times \dfrac{5}{9}$	°F	$\left(°C \times \dfrac{9}{5}\right) + 32$
Plane angle	Radian	0.0174532 degrees $\left(\dfrac{\pi}{180}\right)$	degree	57.29578 $\left(\dfrac{180}{\pi}\right)$
Volume rate of flow	m³/sec	35.31 ft³/sec	ft³/sec (cusec)	0.02832 m³/sec
Fuel consumption	l/km	0.3540 gal/mile	gal/mile	2.825 l/km
Energy	J (Joule)	0.7376 ft/lbf	ft/lbf	1.356 J
Power	kW	1.341 hp	hp (horsepower)	745.7 W (J/sec) 0.7457 kW

SI (metric) units – multiples and submultiples

Prefix	Symbol	Factor by which the unit is multiplied		Example
tera	T	10^{12}	= 1000 000 000 000	
giga	G	10^{9}	= 1 000 000 000	gigahertz (GHz)
mega	M	10^{6}	= 1 000 000	meganewton (MN)
kilo	k	10^{3}	= 1 000	kilonewton (kN)
hecto	h	10^{2}	= 100	hectare (ha = 100 m × 100 m)
deca	da	10^{1}	= 10	
deci	d	10^{-1}	= 0.1	
centi	c	10^{-2}	= 0.01	
milli	m	10^{-3}	= 0.001	
micro	μ	10^{-6}	= 0.000 001	micrometre or micron (μm)
nano	n	10^{-9}	= 0.000 000 001	nanometre (nm)
pico	p	10^{-12}	= 0.000 000 000 001	picofarad
femto	f	10^{-16}		
atto	a	10^{-18}		

BUILDING MATERIALS

Mass

	kN/m²	kN/m³		kN/m²	kN/m³
Aluminium roof sheeting 1.2 mm thick	0.04		*Glass Fibre*		
			Thermal insulation, per 25 mm thick	0.005	
Asbestos cement sheeting			Acoustic insulation, per 25 mm thick	0.01	
Corrugated 6.3 mm thick as laid	0.16				
Flat 6.3 mm thick as laid	0.11		*Glazing, Patent*		
			6.3 mm Glass:		
Asphalt			Lead covered bars at 610 mm centres	0.29	
Roofing, 2 layers, 19 mm thick	0.41		Aluminium alloy bars at 610 mm centres	0.19	
25 mm thick	0.58		Lead, sheet per 3 mm thick	0.34	
Bitumen, built up felt roofing			*Plaster*		
3 layers including chippings	0.29		Gypsum 12.5 mm thick	0.22	
Blockwork, excludes weight of mortar			*Plasterboard Gypsum*		
Concrete, solid, per 25 mm	0.54		9.5 mm thick	0.08	
Concrete, hollow, per 25 mm	0.34		12.5 mm thick	0.11	
Lightweight, solid, per 25 mm	0.32		19.0 mm thick	0.17	
Brickwork, excludes weight of mortar			*Roof Boarding*		
Clay, solid, per 25 mm thick	0.45		Softwood rough sawn 19 mm thick	0.10	
Low density	0.49		Softwood rough sawn 25 mm thick	0.12	
Medium density	0.54		Softwood rough sawn 32 mm thick	0.14	
High density	0.58				
			Rendering		
Clay, perforated, per 25 mm thick			Portland cement : sand, 1:3 mix, 12.5 mm thick	0.29	
Low density 25% voids	0.38				
15% voids	0.42		*Screeding*		
Medium density 25%	0.40		Portland cement:sand, 1:3 mix, 12.5 mm thick	0.29	
15% voids	0.46		Concrete, per 25 mm thick	0.58	
High density 25% voids	0.44		Lightweight, per 25 mm thick	0.32	
15% voids	0.48		*Steel*		77.22
			Steel Roof Sheeting		
Boards			0.70 mm thick (as laid)	0.07	
Cork, compressed, per 25 mm thick	0.07		1.20 mm thick (as laid)	0.12	
Fibre insulating, per 25 mm thick	0.07				
Laminated blockboard, per 25 mm thick	0.11		*Tiling, Roof*		
Plywood, 12.7 mm thick	0.09		Clay or concrete, plain, laid to 100 mm gauge	0.62–0.70	
Concrete, reinforced, 2% steel		23.55	Concrete, interlocking, single lamp	0.48–0.55	
Glass			*Tiling, Floor*		
Clear float, 4 mm	0.09		Asphalt 3 mm thick	0.06	
6 mm	0.14				

BUILDING MATERIALS

Mass

	kN/m²	kN/m³		kN/m³
Clay 12.5 mm thick	0.27		Aluminium, cast	27.50
Cork, compressed 6.5 mm thick	0.025		Brass, cast	87.00
PVC, flexible 2.0 mm thick	0.035		Brass, rolled	83.84
Concrete 16 mm thick	0.38		Bronze	82.27
			Copper, cast	90.00
Timber			Copper, rolled	87.60
Softwoods – Pine, Spruce,			Iron, cast	72.00
Douglas Fir		4.72	Iron, wrought	76.80
Redwood		5.50	Lead, cast	111.13
Pitchpine		6.60	Lead, sheet	111.42
Hardwood – Teak, Oak		7.07	Nickel, monel metal	89.00
Woodwool slabs, per 25 mm thick	0.15		Steel, cast	77.22
			Steel, rolled	77.22
Ashes, coal		7.05	Tin, cast	72.80
Asphalt, paving		22.64	Tin, rolled	72.52
Ballast, gravel		19.20	Zinc	68.60
Brick		20.00		
Cement, potland loose		14.11	Pitch	11.50
Cement, mortar		16.46	Plaster	15.09
Clay, damp, plastic		17.54	Plaster of Paris, set	12.54
Concrete, breeze		15.09	Sand, dry	16.00
Concrete, brick		18.82	Sand, wet	20.00
Concrete, stone		22.64	Slate	29.00
Earth, dry, loose		11.30	Flint	25.90
Earth, moist, packed		15.09	Granite	31.00
Earth, dry, rammed		17.54	Limestone	28.00
Glass, plate		27.34	Macadam	23.57
Glass, sheet		24.50	Marble	28.00
Gravel		18.82	Sandstone	25.00
Lime mortar		16.17		
			Tar	12.00
Masonry, artificial stone		22.60	Terra-cotta	17.90
Masonry, freestone, dressed		25.00		
Masonry, freestone, rubble		21.95		
Masonry, granite, dressed		31.00		
Masonry, granite, rubble		24.30		

PACKAGED MATERIALS

Mass

	kN/m³		kN/m³
Cereals etc.		Oils, in barrels	5.65
Barley, in bags	5.65	Oils, in drums	7.07
Barley, in bulk	6.28	Paper, printing	6.28
Flour, in bags	7.07	Paper, writing	9.42
Hay, in bales, compressed	3.77	Petrol	6.59
Hay, not compressed	2.20	Plaster, in barrels	8.32
Oats, in bags	4.24	Potash	32.14
Oats in bulk	5.02	Red lead, dry	20.72
Potatoes, piled	7.07	Rosin, in barrels	7.54
Straw, in bales compressed	2.98	Rubber	9.42
Wheat, in bags	6.12	Saltpetre	10.52
Wheat, in bulk	8.50	Screw nails, in packages	15.70
		Soda ash, in barrels	9.73
Miscellaneous		Soda, caustic, in drums	13.82
Bleach, in barrels	5.02	Snow, freshly fallen	0.94
Cement, in bags	13.19	Snow, wet, compact	3.14
Cement, in barrels	11.46	Starch, in barrels	3.93
Clay, china, kaolin	21.67	Sulphuric acid	9.42
Clay, potters, dry	18.84	Tin, sheet, in boxes	43.65
Coal, loose	8.79	Water, fresh	9.81
Coke, loose	4.71	Water, sea	10.05
Crockery, in crates	6.28	White lead, dry	13.50
Glass, in crates	9.42	White lead paste, in drums	27.32
Glycerine, in cases	8.16	Wire, in coils	11.62
Ironmongery, in packages	8.79		
Leather, in bundles	2.51		
Leather, hides, compressed	3.61		
Lime, in barrels	7.85		
Oils, in bulk	8.79		

ANGLE OF INTERNAL FRICTION AND MASS OF MATERIALS

Material	Mass in kN/m³	Angle of internal friction°
Ashes	6.3 – 11.6	20 – 40°
Cement	13.4 – 16.8	20°
Cement clinker	14.0 – 16.0	30 – 35°
Chalk (in lumps)	11.0 – 22.0	35° – 45°
Clay		
in lumps	11.0	30°
dry	18.8 – 22.0	30°
moist	20.4 – 25.1	45°
wet	20.4 – 25.1	15°
Clinker	10.0 – 15.0	30 – 40°
Coal (in lumps)	8.0 – 19.0	20 – 45°
Coke	4.0 – 6.0	30°
Copper ore	25.1 – 29.2	35°
Crushed brick	12.6 – 21.8	35° – 40°
Crushed stone	17.3 – 20.4	35° – 40°
Granite	17.3 – 31.0	35° – 40°
Gravel (clean)	14.1 – 20.0	35° – 40°
Gravel (with sand)	15.7 – 19.2	25° – 30°
Haematite iron ore	36.1	35°
Lead ore	50.0 – 52.0	35°
Limestones	12.6 – 18.8	35° – 45°
Magnetite iron ore	40.0	35°
Manganese ore	25.1 – 28.8	35°
Mud	16.5 – 22.8	0°
Rubblestone	17.3 – 19.8	45°
Salt	7.7 – 9.6	30°
Sand		
dry	15.7 – 18.8	30° – 35°
moist	18.1 – 19.6	35°
wet	18.1 – 20.4	25° – 30°
Sandstones	12.6 – 25.0	35° – 45°
Shale	14.1 – 19.8	30° – 35°
Shingle	14.1 – 17.3	30° – 40°
Slag	14.1 – 24.8	35°

Material	Mass in kN/m³	Angle of internal friction°
Vegetable earth		
dry	14.1 – 15.7	30°
moist	15.7 – 17.3	45° – 50°
wet	17.3 – 18.8	15°
Zinc ore	25.1 – 28.3	35°

All materials should be tested under appropriate conditions prior to use in final design.

VALUES OF K_a (COEFFICIENT OF ACTIVE PRESSURE) FOR COHESIONLESS MATERIALS

This table may be used to determine the horizontal pressure exerted by material, p_a, in kN/m².

p_a = mass × depth of material × K_a

Values of δ	Values of K_a for values of Ø				
	25°	30°	35°	40°	45°
0°	0.41	0.33	0.27	0.22	0.17
10°	0.37	0.31	0.25	0.20	0.16
20°	0.34	0.28	0.23	0.19	0.15
30°	–	0.26	0.21	0.17	0.14

The effect of wall friction δ on active pressures is small and is usually ignored.
The above values of K_a assume vertical walls with horizontal ground surface.

Note: The above data should *not* be used in the design calculations for silos, bins, bunkers and hoppers.

APPROXIMATE MASS OF FLOORS

Reinforced concrete floors

	Mass in kN/m²	
Thickness	Dense concrete	Lightweight concrete
100	2.35	1.76
125	2.94	2.20
150	3.53	2..64
175	4.11	3.08
200	4.70	3.52
225	5.23	3.96
250	5.88	4.40

Dense concrete is assumed to have natural aggregates and 2% reinforcement with a mass of 2400 kg/m³.

Lightweight concrete is assumed to have a mass of 1800 kg/m³.

Steel floors

Durbar non-slip		Open steel flooring		
Thickness on plain mm	Mass in kN/m²	Thickness mm	Mass in kN/m²	
			Light	Heavy
4.5	0.37	20	0.29	0.38
6.0	0.49	25	0.38	0.46
8.0	0.64	30	0.44	0.56
10.0	0.80	40	0.60	0.74
12.5	0.99	50	0.74	0.90

Open steel floors are available from various manufacturers to particular patterns and strengths.

The above average figures are for guidance in preliminary design. Manufacturers' data should always be used for final design.

Timber Floors.
Solid timber, joist sizes, mm. Mass in kN/m²

Joist Centres	Decking	75 × 50	100 × 50	150 × 50	200 × 50	225 × 50	275 × 50
400 mm	19 mm Softwood	0.16	0.18	0.21	0.25	0.27	0.30
	19 mm Chipboard	0.19	0.21	0.24	0.28	0.30	0.33
	22 mm Chipboard	0.21	0.23	0.26	0.30	0.32	0.35
600 mm	19 mm Softwood	0.14	0.16	0.18	0.20	0.21	0.24
	19 mm Chipboard	0.17	0.19	0.21	0.23	0.24	0.27
	22 mm Chipboard	0.19	0.21	0.23	0.25	0.26	0.29

The solid timber joists are based on a density of 5.5kN/m³.

WALLS AND PARTITIONS – MASS

Walls

Construction	kN/m²		
	Brick	Block	Brick + Block
102.5 mm thick			
Plain	2.17	1.37	
Plastered one side	2.39	1.59	
Plastered both sides	2.61	1.81	
215 mm thick			
Plain	4.59	2.99	3.79
Plastered one side	4.81	3.21	4.01
Plastered both sides	5.03	3.43	4.23
255 mm Cavity wall			
Plain	4.34	2.74	3.54
Plastered one side	4.56	2.96	3.76
Plastered both sides	4.78	3.18	3.98

Assumed mass of brickwork 21.2 kN/m³

Assumed mass of blockwork 13.3 kN/m³

WALLS AND PARTITIONS – MASS

Partitions

Timber partition (12.5 mm plasterboard each side)	0.25
Studding with lath and plaster	0.76

For specific types and makes of walls and partitions, reference should be made to the manufacturers' publications.

AREAS AND VOLUMES

Areas

Parallelogram	= base × perpendicular height
Triangle	= base × ½ perpendicular height
Trapezoid	= ½ sum of parallel sides × perpendicular height
Circle	= .7854 × square of diameter
Sector of circle	= length of arc × ½ radius
Parabola	= base × ⅔ height
Ellipse	= long diameter × short diameter × .7854
Regular polygon	= sum of sides × ½ perpendicular distance from centre to sides
Surface of sphere	= π × square of diameter
Surface of cone	= area of base + (circumference of base × ½ slant height)

Volumes

Prism	= area of base × height
Pyramid or cone	= area of base × ⅓ height
Sphere	= 4.1888 × radius²

Positions of centre of gravity

Triangle	= ⅓ perpendicular height from base
Parabola	= ⅔ height from base
Pyramid or cone	= ¼ height from base

Side of square of equal area to circle = diameter × .8862
Diameter of circle of equal area to square = side × 1.1284
Circumference of circle = π × diameter

METRIC EQUIVALENTS OF STANDARD WIRE GAUGES

Standard wire gauge	Dia mm	Standard wire gauge	Dia mm	Standard wire gauge	Dia mm
4/0	10.16	3	6.40	9	3.66
3/0	9.45	4	5.89	10	3.25
2/0	8.84	5	5.39	11	2.95
1/0	8.23	6	4.88	12	2.64
1	7.62	7	4.47	13	2.34
2	7.01	8	4.06	14	2.03

THE GREEK ALPHABET

Name	Capital Letter	Small Letter	English Equivalent	Name	Capital Letter	Small Letter	English Equivalent
Alpha	A	α	a	Nu	N	ν	n
Beta	B	β	b	Xi	Ξ	ξ	x
Gamma	Γ	γ	g	Omicron	O	o	short 0
Delta	Δ	δ	d	Pi	Π	π	p
Epsilon	E	ε	short e	Rho	P	ρ	rh
Zeta	Z	ζ	z	Sigma	Σ	σ	s
Eta	H	η	long e	Tau	T	τ	t
Theta	Θ	θ	th	Upsilon	Y	υ	u
Iota	I	ι	i	Phi	Φ	φ	ph
Kappa	K	κ	k	Chi	X	χ	ch
Lambda	Λ	λ	l	Psi	Ψ	ψ	ps
Mu	M	μ	m	Omega	Ω	ω	long 0

CIRCULAR ARCS

The following formulae may be used for exact geometrical calculations.

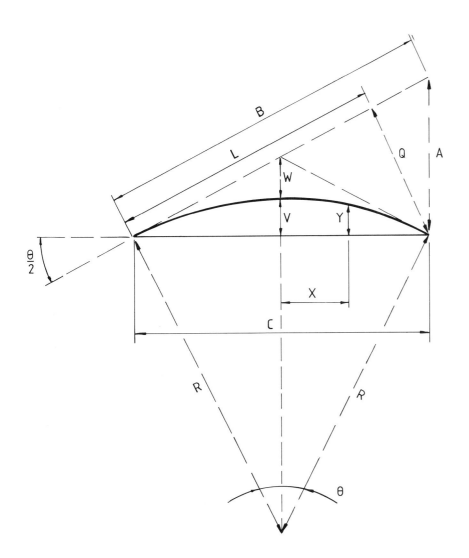

For	Expressions			
N ratio	$\dfrac{1}{\theta} - \dfrac{\theta}{12} - \dfrac{\theta^3}{72C} - \dfrac{\theta^5}{30240}$ etc. (θ Radians)			$\sqrt{\dfrac{R^2}{C^2} - \dfrac{1}{4}}$
θ length	$\dfrac{R}{2L}$ $\Sigma i\,\theta = \dfrac{T}{R}$ $R \times \theta$ radians	$\sqrt{\dfrac{R}{2Q} - \dfrac{1}{4}}$ $\cos\theta = 1 - \dfrac{C^2}{2R^2}$	$\dfrac{T}{2Q}$ $\cos\theta = \sqrt{\dfrac{R^2 - T^2}{R}}$	$\dfrac{R \pm \sqrt{R^2 - T^2}}{2T}$ $\tan\theta = \dfrac{T}{\sqrt{R^2 - T^2}}$
C chord length	$\dfrac{R}{\sqrt{N^2 + \tfrac{1}{4}}}$	$\sqrt{2RQ}$	$\sqrt{T^2 + Q^2}$	
T	$\dfrac{RN}{N^2 + \tfrac{1}{2}}$	$2QN$	$\sqrt{2RQ - Q^2}$	$\sqrt{E^2 - Q^2}$
Q	$\dfrac{R}{2(N^2 + \tfrac{1}{4})}$	$\dfrac{C^2}{2R}$	$R - \sqrt{R^2 - T^2}$	$\dfrac{T}{2N}$
R radius	$\sqrt{C^2 - T^2}$ $C\sqrt{N^2 + \tfrac{1}{2}}$	$2Q\,(N^2 + \tfrac{1}{2})$	$2LN$	$\dfrac{NW}{\sqrt{N^2 + \tfrac{1}{4}} - N}$
V versine	$\dfrac{C^2}{2Q}$ $R - \tfrac{1}{2}\sqrt{4R^2 - C^2}$	$\dfrac{T^2 + Q^2}{2Q}$ $R - CN$	$\dfrac{T(N^2 + \tfrac{1}{4})}{N}$ $R\left(1 - \dfrac{N}{\sqrt{N^2 + \tfrac{1}{4}}}\right)$	$\dfrac{C^2}{8V} + \dfrac{V}{2}$
L	$\dfrac{R}{2N}$	$\dfrac{T}{2} + \dfrac{Q^2}{2T}$		
W	$R\left(\dfrac{\sqrt{N^2 + \tfrac{1}{4}} - 1}{N}\right)$			
W + V	$\dfrac{L}{2\sqrt{N^2 + \tfrac{1}{4}}}$			
A	$\dfrac{QC}{T}$	$2\,(W + V)$		
B	$\dfrac{C^2}{T}$			
Y	$Y - R + \sqrt{R^2 - X}$			

Worked example

Question

A beam is 20 m long and is to be cambered to a circular vertical curve of radius 60 m.

Find

(a)   vertical offset at mid-length
(b)   vertical offset at $\frac{1}{4}$ points
(c)   slope of beam at ends
(d)   true length of beam

Answer

(a)   offset at mid length (or versine)

$$v = R - \frac{1}{2}\sqrt{4R^2 - C^2}$$
$$= 60 - \frac{1}{2}\sqrt{4 \times 60^2 - 20^2} = 0.839 \text{ m}$$

(b)   At $\frac{1}{4}$ point

$$X = \frac{C}{4} = \frac{20}{4} = 5.000 \text{ m}$$

$$y = v - R + \sqrt{R^2 - X^2} = 0.839 - 60 + \sqrt{60^2 - 5^2}$$
$$= 0.630 \text{ m}$$

(c)   Slope of beam at ends

$$\text{ws } \theta = 1 - \frac{C^2}{2R^2} = 1 - \frac{20^2}{2 \times 60^2} = 0.94444$$

$$\therefore \theta = 19.188° \text{ or } 0.3349 \text{ radians}$$

$$\text{Slope at ends} = \frac{\theta}{2} = 9.594° \text{ or } 0.1675 \text{ radians}$$

(d)   Arc length $= R \times \theta$ radius

$$= 60 \times 0.3449 \text{ radians} = 20.094 \text{ m}$$

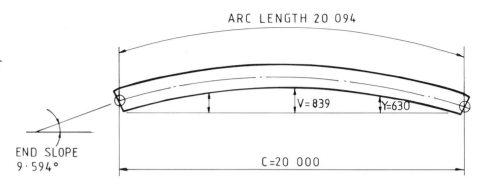

ARC LENGTH 20 094
V = 839   Y=630
C =20 000
END SLOPE 9·594°

CIRCULAR ARCS – LARGE RADIUS TO CHORD RATIOS

The following simplifed formulae are approximate but are usually sufficiently accurate, typically when

$$\frac{R}{C} > 5 \text{ or } \frac{C}{V} > 40 \text{ or } \frac{X}{Y} > 20$$

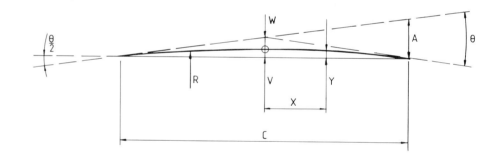

$$V = \frac{C^2}{8R} = W$$

$$A = 4V = \frac{C^2}{2R}$$

$$Y = V\left(1 - \frac{4X^2}{C^2}\right)$$

$$R = \frac{C^2}{8V}$$

$$\frac{\theta}{2} = \frac{C}{2R} = \frac{4V}{C}$$

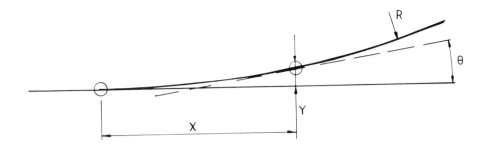

$$Y = \frac{X^2}{2R} \qquad \theta = \frac{X}{R}$$

PRECAMBER FOR A SIMPLY SUPPORTED BEAM

The following formulae can be used to provide deflection and slope values for a beam of uniform stiffness which is uniformly loaded. This enables a precise precamber shape to be determined so as to counteract deflection. The shape will generally be suitable for beams which are not loaded uniformly. Often a circular or parabolic profile is adopted in practice, and is sufficiently accurate.

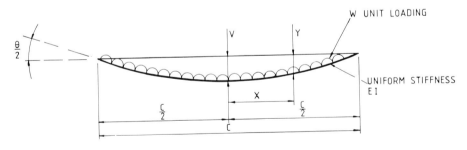

DEFLECTED FORM

Central deflection:

$$V = \frac{5}{384} \frac{WC^4}{EI}$$

Rotation at ends:

$$\frac{\theta}{2} = \frac{WC^3}{24EI}$$

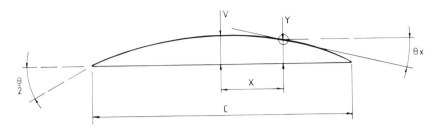

PRECAMBERED FORM TO COUNTERACT DEFLECTION

Precamber at any point:

$$Y = V \left(1 - 4.8 \left(\frac{X}{C} \right)^2 + 3.2 \left(\frac{X}{C} \right)^4 \right)$$

Slope at any point:

$$\theta x = \frac{\theta}{2} \left(\frac{3X}{C} - 4 \left(\frac{X}{C} \right)^3 \right)$$

PARABOLIC ARCS

The following formulae may be used for calculations of parabolic arcs which are often used for precambering of beams.

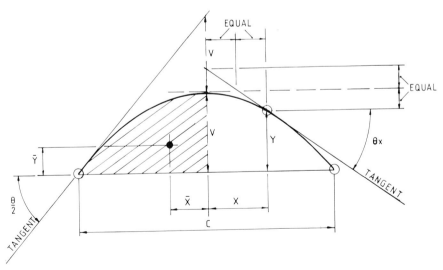

$$\frac{\theta}{2} = \frac{4V}{C}$$

$$Y = V\left(1 - \frac{4X^2}{C^2}\right)$$

$$\theta x = \frac{8VX}{C^2}$$

Approximate arc length =

$$2\sqrt{\left(\frac{C}{2}\right)^2 + \frac{4}{3}V^2} \qquad \text{where } \frac{V}{C} < 0.05$$

For shaded area under curve:

$$\text{Area} = \frac{2}{3} \times \left(\frac{C}{2} \times V\right)$$

$$\overline{X} = 0.375 \times \left(\frac{C}{2}\right)$$

$$\overline{Y} = 0.4 \times V$$

BRACED FRAME GEOMETRY

Given	To find	Formula
bpw	f	$\sqrt{(b + p)^2 + w^2}$
bw	m	$\sqrt{b^2 \div w^2}$
bp	d	$b^2 \div (2b + p)$
bp	e	$b(b + p) \div (2b + p)$
bfp	a	$bf \div (2b + p)$
bmp	c	$bm \div (2b + p)$
bpw	h	$bw \div (2b + p)$
afw	h	$aw \div f$
cmw	h	$cw \div m$

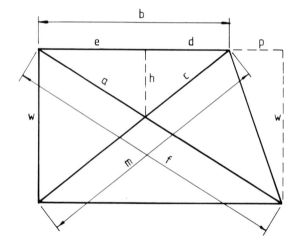

Given	To find	Formula
bpw	f	$\sqrt{(b + p)^2 + w^2}$
bnw	m	$\sqrt{(b - n)^2 + w^2}$
bnp	d	$b(b - n) \div (2b + p - n)$
bnp	e	$b(b + p) \div (2b + p - n)$
bfnp	a	$bf \div (2b + p - n)$
bmnp	c	$bm \div (2b + p - n)$
bnpw	h	$bw \div (2b + p - n)$
afw	h	$aw \div f$
cmw	h	$cw \div m$

Given	To find	Formula
bpw	f	$\sqrt{(b + p)^2 + w^2}$
bkv	m	$\sqrt{(b + k)^2 + v^2}$
bkpvw	d	$bw(b + k) \div [v(b + p) + w(b + k)]$
bkpvw	e	$bv(b + p) \div [v(b + p) + w(b + k)]$
bfkpvw	a	$fbv \div [v(b + p) + (w(b + k)]$
bkmpvw	c	$bmw \div [v(b + p) + w(b + k)]$
bkpvw	h	$bvw \div [v(b + p) + w(b + k)]$
afw	h	$aw \div f$
cmv	h	$cw \div m$

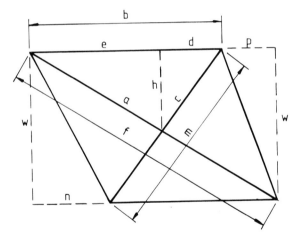

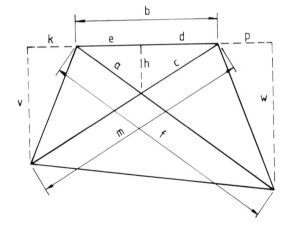

Parallel bracing

$k = (\log B - \log T) \div$ no. of panels. Constant k plus the logarithm of any line equals the log of the corresponding line in the next panel below.

$a = TH \div (T + e + p)$
$b = TH \div (T + e + p)$
$c = \sqrt{(\tfrac{1}{2} T + \tfrac{1}{2} e)^2 + a^2}$
$d = ce \div (T + e)$

$\log e = k + \log T$
$\log f = k + \log a$
$\log g = k + \log b$
$\log m = k + \log c$
$\log n = k + \log d$
$\log p = k + \log e$

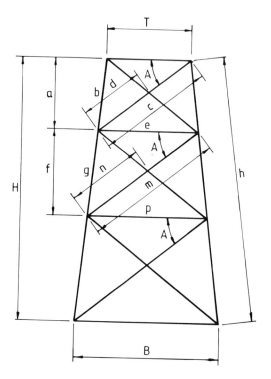

The above method can be used for any number of panels.
In the formulas for 'a' and 'b' the sum in parenthesis, which in the case shown is (T + e + p), is always composed of all the horizontal distances except the base.

INDEX